AF684894

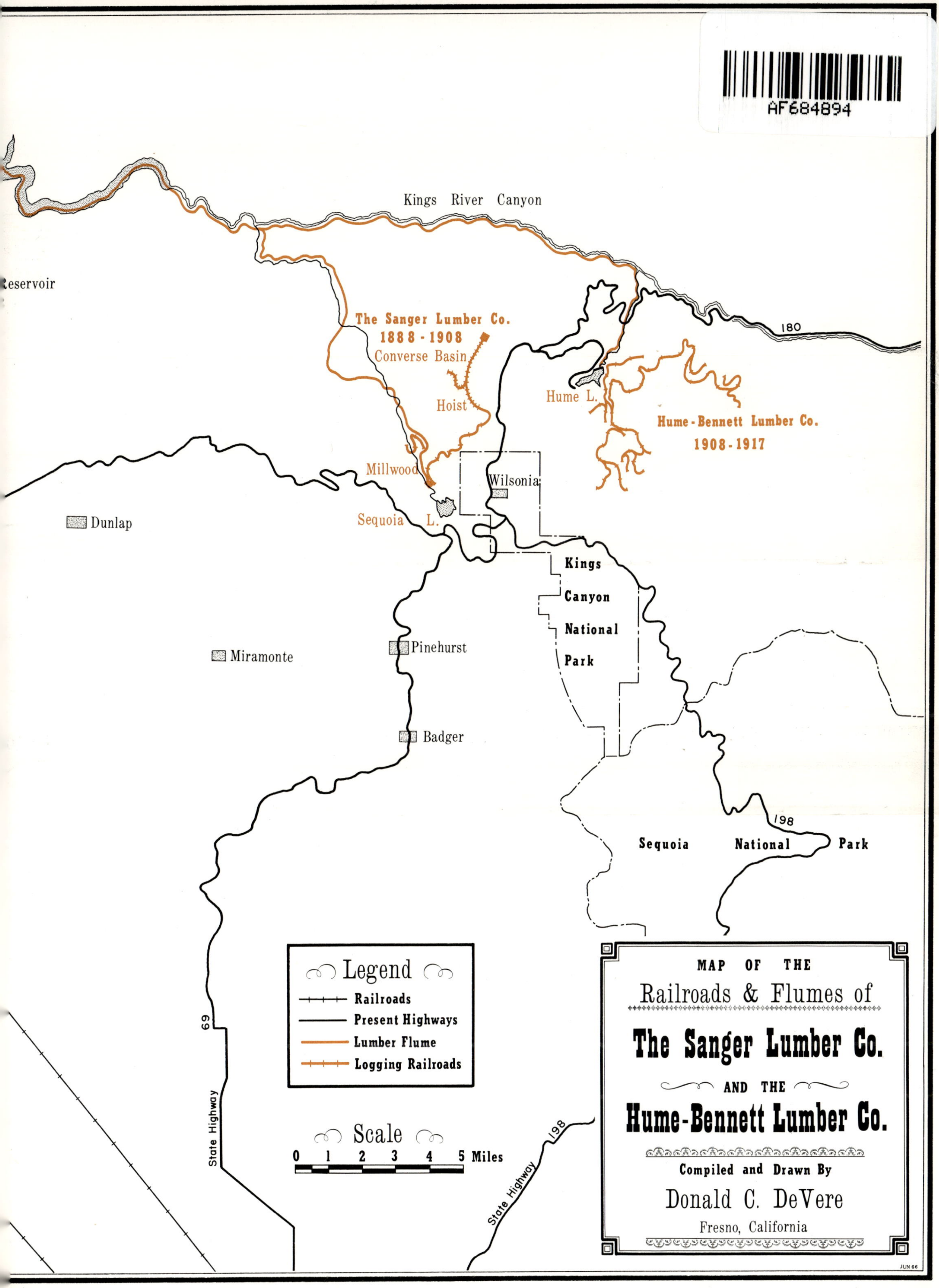
Kings River Canyon
Reservoir
The Sanger Lumber Co.
1888 - 1908
Converse Basin
Hoist
Millwood
Sequoia L.
Hume L.
Hume-Bennett Lumber Co.
1908 - 1917
180
Wilsonia
Dunlap
Kings Canyon National Park
Miramonte
Pinehurst
Badger
198
Sequoia National Park
69
State Highway
State Highway 198
Legend
Railroads
Present Highways
Lumber Flume
Logging Railroads
Scale
0 1 2 3 4 5 Miles
MAP OF THE
Railroads & Flumes of
The Sanger Lumber Co.
AND THE
Hume-Bennett Lumber Co.
Compiled and Drawn By
Donald C. DeVere
Fresno, California
JUN 66

SECTION OF A BIG TREE.

THEY FELLED THE REDWOODS

"Old? They were old when the world was young!"

THEY FELLED THE REDWOODS

By HANK JOHNSTON

A Saga of Flumes and Rails in the High Sierra

TRANS-ANGLO BOOKS
Glendale, California

They Felled The Redwoods

©1983 by Interurbans Publications
All Rights Reserved

No Part of this book may be used or reproduced without written permission from the publisher, except in the case of brief quotations used in reviews.

Library of Congress Catalog Number: 78-104874
ISBN: 87046-003-X
Sixth Printing: March, 1983

Book Design by the Author

Printed and Bound in the United States of America

Published by
Trans-Anglo Books
A Division of
Interurban Press
P.O. Box 6444 • Glendale, California 91205

Front Cover Photograph:
Felling the Mark Twain tree in 1891. *Harold G. Schutt Collection*

Back Cover Photograph
Hume-Bennett locomotive Number One has its gears checked at the engine house at Hume Lake, circa 1912. *Author's Collection*

Frontispiece:
Undercut in the Mark Twain Tree (See page 21.)

FOREWORD

INTRODUCTORY REMARKS

JOHN MUIR said it best: "The Big Tree is nature's forest masterpiece, and so far as I know, the greatest of living things. It belongs to an ancient stock . . . and has a strange air of other days about it, a thoroughbred look inherited from the long ago — the auld lang syne of trees."

Muir was writing, of course, about the mighty Sierra redwoods — unquestionably the King of Trees.

This book is basically a pictorial recounting of the only full-scale lumbering ever conducted of these great trees. It is a tale which seems almost unbelievable when examined in the light of present-day ideas, but unfortunately, it is all true.

The story begins in the late 1880's, and it essentially ends during World War One, but in the short period between, a despoliation of the big trees occurred which was a real national tragedy. It included the ravaging of the Converse Basin — perhaps the finest and purest redwood forest of all.

I must confess that this book was undertaken with a strong feeling of ambivalence. On the one hand, I could not help but be abhorred by the useless profanation of the big trees, yet at the same time, I was grudgingly beset with admiration for the skill, ingenuity and determination of the men who carried out the denudation of the forests. It takes a special brand of audacity to tackle a tree 30-feet across, 250-feet high, and weighing 1500 tons or more, using only a hand saw and an axe!

I therefore ask that the reader be gentle in his condemnation of the workaday loggers herein described. Conservation was not nearly so sacrosanct a word three generations ago. If blame for the cutting of the redwoods must be fixed, let it fall on an apathetic government, a few greedy lumber barons and an indifferent public. And then thank God for men like John Muir, Gifford Pinchot, Stephen Mather, George W. Stewart and all the others who headed the fight to establish our National Parks and Forests. Their cries, like voices in the night, went largely unheeded until it was almost too late.

* * *

At the risk of offending the purists among the reading audience, certain nouns are used in verb form within this text, because they seem to me, more than any regular verb which came readily to mind, to depict graphically the action taking place. I have particular reference to words like "chuted" and "flumed" which, incidentally, were in practical use by those involved at the time under discussion.

Virtually all of the material hereinafter presented was obtained from primary sources such as government records, newspaper stories, company files and personal interviews. Extrapolation has been kept to a minimum. For those who are interested in exact sources, a complete bibliography is listed elsewhere in these pages.

* * *

It seems impossible to convey adequately my appreciation and thanks to the many who worked so hard on this project. If the end result has any lasting merit, it is largely because of their efforts along the way.

The files of the Fresno County Historical Society proved a rich depository of both photographs and information. I am particularly grateful to my good friends at the society, Edwin M. Eaton, president; Dan R. Pollard, curator; and June English, librarian, who so willingly responded to my urgent requests on frequent occasions. Dr. Raymond F. Wood, reference librarian of Fresno State College and a society director, was also helpful.

Other public institutions which provided valuable aid include the Bancroft Library (John Barr Tompkins); the California State Library (Allan R. Ottley); the Fresno County Free Library (Alice Reilly), and the Hackley Public Library of Muskegon, Michigan (Florence Harnau).

Gratitude must be expressed to those who furnished personal recollections from either their own experiences or those of close relatives. They are: Marion Grosse, Mrs. Annie Perry, Mrs. O. A. Olsen, John Cummings, Ross Phillips Jr., Grover E. Johnson, Frank Sayre, the Stafford brothers — Ira, Vern and Dewey, Hume Larsen and especially, Jack Clark, former mill superintendent at Hume Lake.

John Labbe, himself a logging-book author, kindly made the results of his extensive research available. Others who similarly assisted include Doug Richter, Kent Stephens, Don Morrill, Al Harder, Al Graves, George Turner, Kenneth Rea, Bert Ward, Victor Johnson, Ken Barnhart, and Ted Wurm.

C. Russell Johnson, president of the Union Lumber Company, dug into old files for data concerning his grandfather, Charles R. Johnson; Charles Annand, grandson of pioneer photographer C. C. Curtis, granted the use of several rare glass plates, and Mrs. Marguerite Eastwood Welch added little-known facts about John Eastwood and the building of the Hume Dam.

I was assisted many times in Fresno by Roy Walker, a pioneer San Joaquin Valley resident. At Visalia, Jack and Marge Gibson spent endless hours helping me to scrutinize old issues of the Visalia *Delta,* and also supplied excellent hospitality and companionship on several occasions.

The maps and drawings appearing throughout this book are the work of Don DeVere of Fresno. Don is a fine craftsman, a persistent researcher and a knowledgeable railroad historian. I want especially to thank and commend him for the finest and most comprehensive maps of this type that I have ever seen.

Dick Challacombe, long-time Converse Basin student, graciously entrusted me with a portion of his valuable photographic collection, and also filled in certain gaps in the story.

E. M. "Ned" Prescott, an important participant in the scrapping of the old logging equipment, was kind enough to submit to several lengthy interviews before his death which occurred while this book was in preparation.

Clyde Simpson of the Security First National Bank made available his organization's historical files and photographic collection.

I am greatly obligated to Stanley Snook, a walking encyclopedia of High Sierra logging and railroad lore; Dick Abbott, a cousin of the Hume family; James D. Hume, youngest son of George Hume, and Spencer Crump, editorial director of Trans-Anglo books, who all furthered the cause of this production in a variety of ways.

Walter Smith, long-time summer ranger at Sequoia National Park and a former Hume employee, placed his entire personal collection of photographs and notes at my disposal. Additionally, he provided guide service and transportation for trips over the old Hume logging railroad where his experience in the woods proved invaluable.

Bill Reid, a true High Sierra devotee, helped with background material, photographs and research. His own knowledge of the subject is vast and his enthusiasm contagious.

Paul Spivey has spent two decades or more assembling facts and pictures about the old Sequoia-area lumbering days. Paul, for many years a United States Forest Service Ranger at Pinehurst, perhaps knows the Sequoia forests better than anyone else. His aid in fixing the exact locations of old logging chutes, railroad spurs and the like, could not have been duplicated elsewhere.

Two persons made this book almost an individual crusade, and became practically fellow authors as a result. Harold G. Schutt, guiding light of the Tulare County Historical Society and editor of its excellent bulletin *Los Tulares,* is an expert on "Sequoiana." He has personally interviewed many of the old lumberjacks over the years, and has assembled a brilliant private collection of historical photographs. I shall be forever in his debt because of his unstinting and invaluable support during all phases of the preparation of this book.

Last, but by no means least, thanks must somehow be properly expressed to Jean Hume Browning, a daughter of George A. Hume. Jean Browning lived many childhood days along the shores of Hume Lake, and later, through her family, acquired certain material which is unique. When I first approached her more than a year ago for assistance in this project, I had little idea what an immense favor I was asking. I do not really know how many hours Jean spent in fulfilling my countless requests for information, as each single one undoubtedly involved days of researching through musty company journals, piles of forgotten correspondence and nearly illegible old records. I do know that her task was an enormous one. The fact that she did this without asking permission to censor the final copy, and with the idea only of sharing her historical material with a larger audience, is forever to her credit. If the reader receives any ultimate benefit from the contents of these pages, he owes it in large measure to Jean Hume Browning!

* * *

Here, then, is *They Felled the Redwoods* — a labor of love, presented with a strong feeling of inadequacy.

The big trees have outlived a thousand catastrophies down through the ages. I am quite sure they will survive this book.

HANK JOHNSTON
Long Beach, California
July 4, 1966

A SYLLABUS

TABLE OF CONTENTS

THE SHAY

▣

"Work Horse of the Woods"

CHAPTER ONE

THE BIGGEST TREES IN THE WORLD

HIGH ALONG the western slopes of California's Sierra Nevada mountains grow the mightiest trees on the face of the earth. These forest giants stand so tall and are so big around that they virtually defy description. Their botanical name is *Sequoia gigantea* but they are variously called redwoods, sequoias or sometimes simply "The Big Trees."

The fame of these primeval monarchs is world wide. They are the largest and among the oldest of living things. Over the years millions of people have journeyed far to see them; countless tales have been told about them. Today these trees are revered and protected in a manner befitting their exalted position as one of nature's wonders.

But it was not always thus.

During the late 19th and early 20th centuries, thousands of these huge redwoods were felled by the cutter's axe in the greatest orgy of destructive lumbering in the history of the world. Nearly 20 per cent of the total number was ruthlessly destroyed by men who acquired the right by fraud and deceit.

"Cut 'er down, boys, there's plenty more over the next hill," was the cry during those earlier days. And cut them down they did — one giant after another in rapid succession. Often the trees were so large and so brittle that they shattered into worthless rubble as they hit the ground. Less than half the felled timber was utilized. The waste was appalling.

It is indeed ironical and perhaps fitting that financial ruin became the only reward for the freebooters who plundered the forests. Even the eastern businessmen who later legitimately bought the remnants of the operation suffered huge losses for their trouble.

The great timberlands of America have been a source of controversy between lumberman and conservationist for almost a century. It is an argument that, like all others, has more than one side. In the matter of the logging of the giant redwoods, the end result can hardly be defended as being in any way desirable; yet the pure physical effort of the doing was in all fairness a marvel of ingenuity that is nearly unmatched

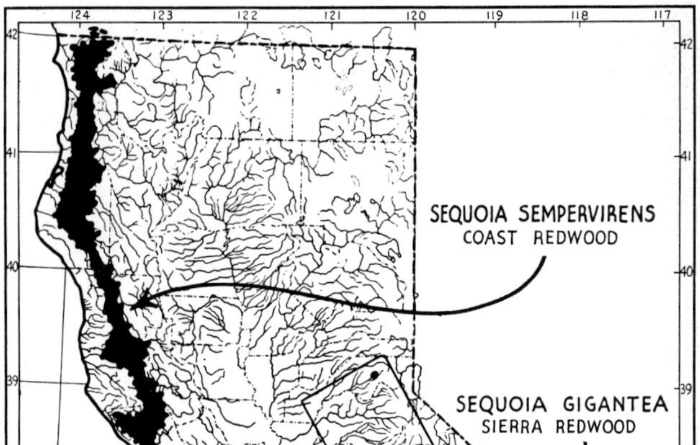

in any other time or place. When viewed from that angle alone and apart from anything else, it is a fascinating example of man's resourcefulness and daring.

The story really begins thousands of years ago when these ageless relics we call the redwoods were born.

* * *

Redwoods are cone-bearing trees belonging to the family *Taxodiaceae* and are related to bald-cypress, pines, firs, cedars and other cone-bearing groups. They fall into three distinct classifications: Sierra redwoods (*Sequoia gigantea*), coast redwoods (*Sequoia sempervirens*) and dawn redwoods (*Metasequoia*). The latter is a deciduous cousin of the others that was believed long extinct until found as a hitherto undetected living tree in China in 1944. Today it is widely cultivated as an interesting ornamental.

Millions of years ago, representatives of these various trees appeared in many parts of the world. Subsequent glaciers and other radical changes in climate and topography of the earth gradually restricted redwoods to their present locations. They are far from a vanishing species, however, as young trees of both main types display phenomenal vigor.

For many years, foresters, botanists and naturalists have turned out a steady stream of literature about the redwoods. Nevertheless, considerable confusion still exists in the public mind concerning the big trees.

"Which trees are redwoods, which are sequoias, and what's the difference?" California visitors often ask. Their perplexity is understandable for even the experts do not completely agree on the matter.

Tradition has long favored the use of "redwood" for the coastal trees (*Sequoia sempervirens*), and sequoia or Big-Tree for the inland variety (*Sequoia gigantea*). But a "big tree" can be any kind, and both types are truly gigantic. We find a "Big Trees Park" and the "Santa Cruz Big Trees" near the coast — both of which feature *Sequoia sempervirens*; similarly, in the Sierra Nevada are Redwood Meadow, Redwood Mountain and several Redwood Creeks — all based on *Sequoia gigantea.*

In was ruled in 1951 by the California attorney general that the designation "redwood" as the official state tree applied to both species without prejudice; the Save-the-Redwoods League which was formed in 1918 fights impartially for the preservation of both *Sequoia gigantea* and *Sequoia sempervirens.* The position that "redwood" applies only to *Sequoia sempervirens* now hardly seems tenable.

In an effort to clarify the situation, the California Division of Forestry some years ago adopted the use of the names coast redwood and Sierra redwood as its recognized nomenclature. This approach, which appears to indicate properly the geographical differences as well as the generic similarities between the two types of trees, has received general acceptance among authorities.

* * *

The earliest recorded discovery of the California redwoods occurred in 1769 when the Portola expedition — a group of early Spanish explorers and settlers — reached the area of Monterey Bay. Fray Juan Crespi, diarist for the group, wrote under the date of October 10, that "the coastal plains and low hills were well forested with very

Here are the coastal redwood trees that are typical of those in northern California's Redwood Empire.
(Spencer Crump Collection)

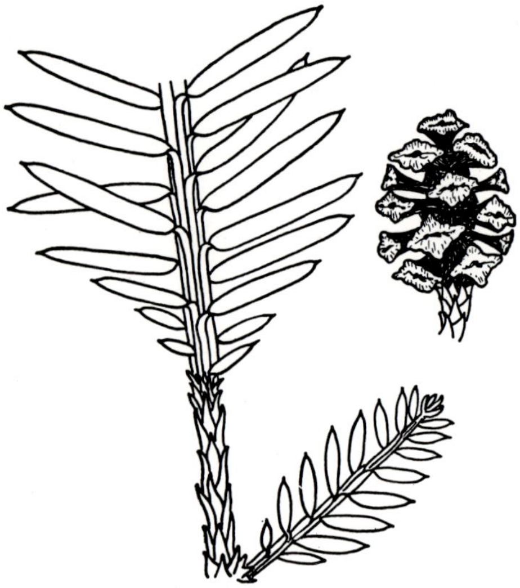

Foliage and Cone (natural size) of Coast Redwood

Foliage (¾ natural size) and Cone (½ natural size) of Sierra Redwood

high trees of a red cedar not known to us . . . in this region there is a great abundance of these trees, and because none of the expedition recognizes them they are named redwood from their color."

The trees were, of course, the coast redwoods (*Sequoia sempervirens*). It seems strange indeed that these giants were not detected during the more than two centuries between the discovery of California by Cabrillo in 1542 and the Portola expedition. It is known that Portuguese, Spanish and English explorers were in the redwood area several times during the period. Somehow they all failed to notice the tall trees. At least no comment about them appears in any expedition report.

Once discovered, the redwoods lay almost untouched for many years. Russian colonists conducted some minor logging along the Russian River in the early 1800's, and a church built of redwood at Fort Ross in 1812 is still standing. But in the main, the unusual properties of the red timber attracted little attention until the great California gold rush of 1849.

Almost overnight, hordes of goldseekers descended on San Francisco. Bridges, barns, houses and stores were hastily thrown up throughout the Mother Lode. Often they burned down just as rapidly. The 49'ers rediscovered what the Russians and Indians before them had known — that redwood was remarkably resistant to fire. Suddenly it was in big demand as a building material. Within a short time, sawmills and lumber camps were operating all along the northern coast.

After untold centuries of battle against every natural hazard, the redwoods now faced the most relentless enemy of all — the logger and his axe.

* * *

Scientific interest in the coast redwoods preceded the gold rush by several decades. In 1824, English botanists classified the trees as members of the bald-cypress species (*Taxodium*). Stephen Endlicher, an Austrian botanist, determined in 1847 that the redwood was not really a bald-cypress at all but a member of an entirely new genus. He named it *Sequoia sempervirens* (always green).

The generic name *Sequoia* is popularly supposed to have been chosen by Endlicher to honor George Guess, inventor of the Cherokee system of writing. Guess, son of a Cherokee woman and a German trader named Nathaniel Gist, had the Indian name of Sequoya (or Sequoyah) and became famous by teaching many of his people to read and write their own language. Certainly Guess deserves the tribute and the story may well be true, but there is no real evidence that Endlicher had ever heard of Sequoya and no such reference appears in Endlicher's own writings on the subject.

The coast redwood is the tallest evergreen tree in the world. Many of them are 350 feet high. The famous Founders Tree in Humboldt County has been measured at 364 feet in height making it probably the tallest single specimen on earth.*

*There is dispute about which tree is actually the tallest. Recent discoveries seem to indicate that at least one, and possibly several, redwood trees may surpass the Founders Tree.

George Guess, the half-breed "Sequoyah," who achieved fame as the inventor of the Cherokee system of writing. In 1821, Guess devised an alphabet of 87 characters, called by some linguists the second best system of writing ever known. The photo above is from a portrait made in 1828. Whether or not Guess' name was given to the Sequoia trees is a matter of conjecture.

The trunks of the coastal trees seldom exceed 20 feet in diameter, but thousands have diameters of from 12 to 16 feet. Since the bole tapers only slightly, the yield of lumber from a single tree is gigantic — in some cases more than 350,000 board feet. The wood is heavy and strong and adapted to a wide variety of uses because of its extreme resistance to fire, insects and rot. About 25 per cent of the lumber cut in California each year comes from the coastal groves, many of which are still owned by private interests.

The coast redwood grows only within the summer fog belt of the California Coast Range from the extreme southwestern corner of Oregon southward to the Santa Lucia Mountains near Monterey. The species thus occurs in a 500-mile-long strip which varies from 10 to 40 miles in width. The average width is about 20 miles. Its finest stands are always in low river valleys and flats but it is found infrequently at an elevation as high as 3000 feet above sea level.

The root system of the coastal tree is broad and shallow. The bark is reddish-brown with a slight grayish cast and is very fibrous. It varies in thickness from a few inches to more than a foot at the base of large trees. The leaves are arranged in a series of flat sprays along both sides of the twigs and resemble expanded needles in appearance. Reproduction takes place by both seeds and root crown sprouts.

The mature coast redwood ranges from 800 to 1500 years of age. A few are thought to be more than 2000 years old.

* * *

In 1852, almost a century after Fray Juan Crespi's historic visit to Monterey, an amazing story trickled down from a remote area of the High Sierra. It told of the discovery of trees that were even larger than the coastal giants — *in all probability the biggest trees in the world!*

The news came from Murphy's Camp, a brawling, rip-roaring gold settlement in Calaveras County populated mainly by miners, gamblers, prostitutes and bandits. A. T. Dowd, one of four hunters employed by some of the more affluent miners to furnish game, was tracking down a bear he had wounded, when he wandered into a section of what is now the Calaveras Big Trees State Park. Dowd found himself face to face with the largest trees that he had ever seen! He quite naturally lost all interest in the errant bear and rushed breathlessly back to camp, full of the news of his remarkable discovery. Unfortunately, no one at Murphy's would believe his tale; many, in fact, thought he was drunk, and not a single person would go back with him to see the trees.

Dowd was obviously a most resourceful individual. He bided his time for what seemed a reasonable period, then rushed into camp again and announced that he had just shot "the largest grizzly bear that I ever saw in my life." The bear, of course, was a fake. But the story was apparently more believable than the one about the trees had been, for this time Dowd was able to persuade a group of miners to accompany him into the woods. On the pretext of searching for the grizzly, he steered the party into the grove of giant trees. As the miners gaped in astonishment, Dowd yelled triumphantly, "Now, boys, do you believe my story?"

Newspapers from San Francisco to England published accounts of the remarkable discovery and Murphy's Camp soon became a mecca for

the curious who wanted a first-hand view of the forest monarchs. For a time they were thought to be the only such trees in existence, but within a decade, almost all of the other Sierra redwood groves were located and at least partially explored.

Others before Dowd had seen the big Sierra trees. The J. R. Walker party made a journal entry about the "red-wood" when passing near the Merced and Tuolumne Groves in 1833. There are unconfirmed reports of two other sightings of the Calaveras Grove prior to 1852. But Dowd's adventure unquestionably was the event that brought the Sierra redwoods to the attention of the world.

English scientists were the first to classify the "new" tree. They placed it in a genus of its own and called it *Wellingtonia gigantea* after the Duke of Wellington who was in the news at the time. American botanists, considering it a species of bald-cypress and being equally patriotic, named it *Taxodium washingtonianum.* The dispute was resolved in 1854 when Joseph Decaisne, a French botanist, determined that the tree was really another member of the genus *Sequoia* and christened it *Sequoia gigantea.* This classification is still in general use despite recent attempts by some botanists to create a separate genus for the big tree called *Sequoiadendron giganteum.*

The Sierra redwood is truly an immense specimen. The General Grant Tree in Kings Canyon National Park has a base diameter of 40 feet and many trees have trunks more than 25 feet thick. The inland redwood is shorter than the coastal tree (seldom more than 300 feet in height), but contains a much greater volume of wood. The General Sherman Tree in Sequoia National Park, called the world's largest tree, consists of 600,000 board feet of lumber — enough to build at least 40 five-room houses. It has been estimated that the weight of the whole tree — trunk, branches, foliage and root system — would fill 205 standard railroad freight cars carrying 30 tons each, making a train one and three-fourths miles long!

Mature Sierra redwoods are often 2500 to 3500 years of age — some perhaps 4000 years

The General Grant Tree in Kings Canyon National Park is generally conceded to be the second largest tree on earth. It was officially designated the "Nation's Christmas Tree" in 1926 by the United States Department of the Interior. In 1956, it was made a National Shrine by an Act of Congress. Each Christmas there are special services held at the tree. (National Park Service)

old. (Imagine a tree still growing that was 200 feet tall at the time of Christ!) They have long been publicized as "nature's oldest living things," but botanists now believe that this honor actually belongs to the bristlecone pine of the Great Basin. Surely the redwoods are next to the oldest.

They have very thick bark — usually from one to two feet thick at the base of the trunk although a thickness of more than four feet is known. The bark is almost asbestos-like in its resistance to fire and insects. Its color is cinnamon red with an occasional purplish sheen. The wood is salmon-pink which becomes dark maroon when exposed to air. It is very brittle, quite soft and decidedly inferior to that of the coast redwood as a building material. It has virtually no commercial use today and almost all groves are completely government controlled.

The leaves of the Sierra redwood are one-fourth to one-half an inch long and look like sharp-pointed scales. The roots do not penetrate downward for much more than six feet which is almost unbelievable for such a giant tree. The lateral spread, however, may reach out as far as 300 feet. Because of this shallow, sensitive root structure, some of the more heavily visited trees have been fenced off for a distance at the base to prevent damage.

Reproduction of the *Sequoia gigantea* is by seed only and almost perfect conditions are required. The coastal trees, which sprout from the stump when cut, are more than 10 times as numerous as the inland giants.

Sierra redwoods occur along the western slope of the Sierra Nevada between Placer County on the north and Tulare County on the south. Within this 250-mile belt, the trees are found in 72 groves having a total area of 35,600 acres. Only eight groves are north of the Kings River so the major distribution is in Tulare and southern Fresno counties. The inland giants grow intermingled with other species — principally white fir and sugar pine. They are found between the extreme elevations of 3000 and 8900 feet; optimum elevations are from 5000 to 7500 feet above sea level.

* * *

The world was very slow in accepting the Sierra redwood as a reality. The trees were just too big to be believed. Many people felt that the tall tales emanating from the Sierra back country were merely the ravings of drunken miners — certainly nothing to be taken seriously by civilized folk.

Contemporary artist's sketch of the felling of a Sierra redwood by drilling a series of holes in the trunk with a long auger. This is probably the tree cut at the Calaveras Grove. (Author's Collection)

But there were enterprising promoters who not only believed but were quick to see that the forest giants had great potential as sideshow displays.

If the public doubted, surely the public would pay to see.

The first great tree to be felled was at the behest of just such a promoter. In the summer of 1853, barely a year after Dowd's discovery, five men spent 22 days boring holes with a long auger in the trunk of a 302-foot specimen at the Calaveras Grove. Wedges were then inserted and driven in by huge logs hung by ropes from the trunk. One day when the cutting crew was safely at lunch, a sudden gust of wind toppled the monster and it fell heavily to the ground breaking its trunk into many pieces. The effect was like that of an earthquake and the shock could be felt for a great distance. The immense weight of even a section of the huge trunk prevented anyone from moving

This sketch shows a section of the big tree that was downed by boring countless auger holes. Note the many ridges made by the bit. (Author's Collection)

the remnants. Instead, the bark was stripped for a distance of 21 feet up the trunk and carefully reassembled for showing in San Francisco and later New York where it vied with a P. T. Barnum extravaganza. At Calaveras, a pavilion was erected over the great stump which soon became a favorite spot for San Francisco cotillion groups. It was reported that 49 persons simultaneously used the "floor" at one event including dancers, orchestra and spectators.

"The laborious vandals had seen the biggest tree," John Muir later decried. "Then, forsooth, they must try to see the biggest stump and dance on it!"

Another promoter, seeing the interest in the first exhibit, conceived the scheme of removing the bark from a standing tree, piece by piece, rather than going to the expense of cutting it down. He carefully accumulated 116 linear feet of bark and shipped it to New York and London for re-assembly on a framework display. The debarked tree was unable to survive without its protective coating and died some years later.

Eventually whole trunk sections of large trees were shown at several world's fairs and expositions including the Centennial Exhibition at Philadelphia in 1876 and the World's Columbian Exposition at Chicago in 1893. C. C. Curtis, a pioneer photographer in the Sequoia area, took a complete set of photographs of the latter cutting and accompanied the tree to the exposition armed with 50,000 prints to sell. When the Chicago concessionaire refused to permit Curtis to ply his trade on the grounds that he had an "exclusive franchise," the photographer was forced to return to California in an almost destitute condition.

Other big trees were cut for private exhibits from time to time. Reports indicate that few, if any, of the various redwood displays achieved particular financial success. The idea that it was all just a "California Hoax" was hard to overcome.

* * *

The first logging of Sierra redwoods occurred in the 1860's. It increased so rapidly that within only a few years agitation against felling the giants began to be heard. In 1873, the California State Legislature passed a law providing that "any person or persons who shall wilfully cut down or strip of its bark any tree 'over sixteen feet in diameter' in the groves of big trees situated in the counties of Fresno, Tulare or Kern or shall destroy any of said trees by fire, shall be guilty of a misdemeanor." The phrase concerning the diameter limit appears in quotation marks in the original statute, as if the author were not really sure that trees of this size actually existed. This law is still on the statute books but there is no record of any action ever being filed under it although many early violations obviously occurred.

A number of small sawmills were built in the southern Sierra in the 1870's and 1880's despite the difficulties involved in transporting lumber from the mountains. Freighting by wagon was very costly — about $6 to $8 a thousand feet. Attempts to use tramways and small flumes met with only limited success.

In 1883, Smith Comstock, a lumberman from the Santa Cruz area, began cutting at Big Stump near General Grant Grove. Some very large redwoods were felled in this six-year operation.

But all these early logging ventures were inconsequential when compared with what was yet to come. There were millions of board feet of timber waiting on the Sierra mountainside — and in the most massive individual chunks that anyone had ever seen!

Two big-time lumbermen from San Francisco coveted that treasure, and they didn't especially care what means they used to get it.

In the middle 1880's they made their move . . .

World's Fair Tree

The most ambitious "big tree" exhibit of all was that shown at the World's Columbian Exposition at Chicago in 1893. The United States Department of the Interior paid $15,000 to a local contractor for the cutting and delivery of a big redwood. The tree selected, the "General Noble," was situated about three miles northwest of General Grant Park. The big tree was first cut off 50 feet above the ground and the stump then hollowed out for a distance. Exterior sections, 14 feet long and about six inches thick in addition to the bark, were carefully sliced apart. These pieces were marked, crated and hauled to the railroad for shipment to Chicago where they were reassembled. Photos at right and below were taken by C. C. Curtis, pioneer Sequoia area photographer, who made up 50,000 prints of the operation to sell at the exposition. When he was not permitted to do so, he was forced to give up the photo business. The stump, called the "Chicago Stump," can still be seen near Converse Basin. (Photos Harold G. Schutt Collection)

Mark Twain Tree

The Mark Twain Tree was an immense Sierra redwood left by early Sequoia area lumbermen. By arrangement with Colis P. Huntington, president of The Southern Pacific, the tree was felled in 1891. Cross sections of the enormous trunk, 17 feet inside the bark, were sent at Huntington's expense to the American Museum of Natural History in New York and the British Museum in London. The exhibits are still in evidence at these two locations. The remarkable photo at left by C. C. Curtis shows the Mark Twain as it starts its downward journey to the ground. Note the deep side cut made so the saw could be handled. Picture at right was taken during the assembling of the trunk section at the New York Museum in 1892. Another Curtis photo below illustrates how 50 men stood on the massive stump. (Photos Harold G. Schutt Collection)

End view of the Mark Twain trunk. (Harold G. Schutt Collection)

CHAPTER TWO

THE KINGS RIVER LUMBER COMPANY

WHEN CALIFORNIA achieved statehood in 1850, most of the timber land was in possession of the federal government. Like other public land, it was open to claim by any citizen under certain rules and conditions. Millions upon millions of forest acres were involved.

Following the Civil War, settlers by the thousands poured into the west seeking a piece of ground and a new way of life. But these early pioneers were not at all interested in the timber. They regarded it as having little value, and sought out mineral property or ranch sites instead.

As a result, when eastern lumber barons turned their attention to California after the completion of the transcontinental railroads, they found that practically the entire timber belt was still in the public domain. They lost little time in using every available means to gain possession. Claims were acquired under the Homestead Law, the Timber and Stone Act, the mining laws and the Swamp Land Act. The public land laws were violated so extensively that they accomplished little else than ridding the people of their natural heritage.

Timber frauds were openly perpetrated. Dummy locators were imported by the score from all over the country to file claims. Once the patent was issued, title was turned over to the lumber operator who had prearranged the deal. The locator was usually paid from $100 to $300 for his services.

Most of the choice timber lands were appropriated before the government awoke to the fact that urgent action was necessary if it wished to preserve any of its timber. On March 3, 1891, a law was passed which enabled the president to withdraw timber lands from the public domain and set them aside as forest reserves. This was the start of America's national forest system which has since played a most important part in the conservation program.

Unfortunately, in many instances it was too late. Great tracts of irreplaceable trees had already been lost to the lumbermen including one of the finest and purest stands of Sierra redwoods.

It all happened before anyone was really aware of what was going on.

In the summers of 1886 and 1887 there was a sudden and unexplained rush of filings on recently surveyed timber lands south of the Kings River in southern Fresno and northern Tulare counties near what is now Kings Canyon National Park. Large numbers of locators appeared almost daily in Visalia to file claims. The Visalia Weekly *Delta* reported on June 16, 1887, that the land offices at both Stockton and Visalia were "overrun with applicants."

It was common knowledge in the San Joaquin Valley that real estate agents in the Visalia area were running "tours" to the nearby timber belts for prospective purchasers who had not yet used their claiming rights. Caravans of jam-packed stages could regularly be observed traveling to and from the Sierra. Meals were arranged for at farm houses along the way. Once in the mountains, vacant quarter sections (160 acres) were chosen from township maps. Many times the locator was not even shown the same section on which he was filing — a legal requirement — but few seemed to know or care. A quick signature back at the land office and the claim was entered. The cost under the Timber and Stone Act of 1878 was $2.50 an acre, and no residence or improvements were required.

Prospective locators were imported from as far away as San Francisco. Agents rounded up idlers, drunks and hangers-on from nearly every valley town. It was reported that all arrangements were being taken care of by unknown parties who were willing to pay the expense of the mountain trip, provide the filing fee and charge, and then kick-in a cash settlement for the locator's trouble. Payments varied from $100 to $600 depending on the particular sections in question. Once the deed was delivered, it was signed over to the agent and the matter was closed as far as the locator was concerned. He had made a nice profit; spent a pleasant, expense-paid day in the mountains, and no harm done. Or so he thought. Actually there was nothing illegal about selling a deed once it had been properly filed and paid for, but prior collusion was another matter.

Rumors were so widespread about the land grabbing that the government sent investigators into the area several times. It was extremely difficult to find the transients who had been placing the bulk of the claims, and even harder to pin them down on exactly what arrangements they had worked under. Eventually complaints were placed against V. D. Knupp, well-known Visalia real estate man, and 34 locators who had filed on choice timber sections near Mill Flat Meadow. Special Agent C. W. Greene of the U. S. Interior Department angrily charged that these parties had fraudulently conspired to acquire 5440 acres of government land by collusion, and that the 34 locators had been paid from $500 to $600 each for their claims. After a preliminary hearing on March 31, 1892, the matter was dropped, apparently for lack of enough evidence.

Hiram C. Smith, partner and general manager of the Kings River Lumber Company. Smith was in direct charge of the logging and fluming operation. (Dick Challacombe Collection)

* * *

The first public knowledge concerning the men behind the Kings River land grab came on March 1, 1888, when the San Francisco *Chronicle* reported that Hiram C. Smith and Austin D. Moore, prominent San Francisco lumber merchants, had recently "purchased thousands of acres of Tulare County land and intend to build two sawmills with a combined capacity of 140,000 board feet a day about 60 miles northeast of Visalia on the Kings River."

Included in the story was the amazing announcement that a "standard gauge railroad will be built to the two mills from Centerville up the Kings River Canyon to the mill sites, and then extended into the timber." Centerville was at the time a leading San Joaquin Valley town. Whoever was responsible for the interesting idea of building a railroad up the Kings River Canyon had apparently never been in that rugged part of the country or he would have realized that such an undertaking was impractical.

The real reason behind the multitude of land claims during the previous two summers was now clear. About 30,000 acres of the finest mountain timber land had been illegally and systematically acquired for the purpose of conducting a gigantic logging enterprise. Included in the holdings were vast stands of fir, pine and cedar as well as some of the very finest Sierra redwood tracts. The cost had been dirt cheap — $2.50 an acre plus the locators' bonus payments and the agents' fees.

Smith and Moore were important figures in the lumber business along the coast. They were the principals in Moore and Smith Lumber of Stockton, and the Port Discovery Mill with 18,000 acres of timber in Washington. Moore, who owned two-thirds of the partnership, was also involved in a number of mining and ranching ventures. His San Francisco home was a showplace. Smith was the more practical lumberman and it was he who assumed direct charge of the Kings River operation. Within a month of the first announcement, Smith had further news of the project.

"We have given up our railroad," he said. "Instead our plan is to build a great flume to bring lumber from the mill sites in the mountains on the south side of the Kings River to a point on the north side in the San Joaquin Valley 50 miles away."

As if to prove that he was serious about the proposed flume, Smith sent a party of seven persons into the Sierra on April 5, 1888, for the avowed purpose of making a preliminary survey of the best route.

"Right-of-way is a problem," Smith observed, "as is the proper grade. We shall feel our way along and decide on a destination for our flume when all factors have been carefully considered."

Indications were that Fresno, the principal valley town, would be the eventual terminus.

Mill Flat Meadow, site of the flume reservoir, as it appeared in 1885. The cattle belonged to a local dairyman. (Fresno County Historical Society)

On April 24, 1888, the new lumber organization was formally incorporated under the name of the Kings River Lumber Company with a capital stock of 10,000 shares, par value $100. Officers were Austin D. Moore, president; Hiram C. Smith, vice-president, and John R. Hanify, secretary. Alonzo Grim and John R. Jarboe were additional directors.

The flume survey was completed in January, 1889. By June, workers began arriving on the overland stage from San Francisco and construction started on the sawmills themselves. Two mills were planned — one above and one below the snow line, so they "can be run winter and summer on a year-round basis." Also proposed was a large reservoir on Mill Flat Creek about a mile above the upper mill to provide a steady supply of water for the flume.

On July 29, 1889, work commenced on the flume itself. Lumber was supplied from the upper mill which began cutting on July 6. The lower

Lower mill of the Kings River Lumber Company. Note flume in foreground. This mill was at a lower elevation than the upper mill and was originally planned to run on a year-round basis but the idea was not carried out. (Bill Reid Collection)

mill, two miles below the upper mill, was not opened until August 26. All materials for the two sawmills were laboriously freighted into the mountains by team and wagon over rudely constructed roads. It was a back-breaking job but only one major accident was reported when a huge gang-edger fell from a wagon. It was repaired at a cost of $700.

Smith anticipated no difficulty in completing the immense project.

"We expect to build about four miles of flume per month on the steep upper end to a place on the Kings River called Trimmer Springs," he confidently stated. "We should make eight miles per month on the lower portion which is almost level. A large force of men has been engaged."

Speculation about the eventual terminus of the flume was now widespread in the San Joaquin Valley. Fresno and Centerville openly coveted the Kings River Lumber Company's business. The new town of Sanger, 15 miles east of Fresno, was also being mentioned as a candidate since it was conveniently situated on the Porterville branch of the Southern Pacific.

"Teams are at work hauling lumber to the lower mill for year-round cutting," the Visalia Weekly *Delta* commented on August 15, 1889. "One and one-half miles of flume are now built. Termination of the flume is still uncertain and if Smith and Moore know, they are not saying. We cannot believe that Fresno will allow the village of Sanger to take away so valuable an enterprise."

Sanger residents had other ideas.

They proposed a donation of 65 acres of land to the Kings River Lumber Company if the flume would be built to their town. On August 29, 1889, Smith and Moore accepted the offer.

"The flume is ours," bragged the Sanger *Herald*. "All right-of-way has been obtained and the matter is closed."

Fresno interests were irate! The idea that an upstart hamlet had beaten them out for such a valuable facility was hard to swallow.

"How can it be?" one Fresnan declared in disgust. "Surely Smith and Moore will reconsider this proposition. Sanger is little more than a railroad siding."

Certainly the town was no metropolis, having but 250 inhabitants. Only a year before it had been a wheat field!

In 1887, a gentleman from Indiana named Joseph Sanger, Jr., came to San Francisco to attend the annual convention of the Railroad Yardmasters Association in his official capacity as that organization's secretary-treasurer. During the meeting he was informed that a group of California railroad officials planned to honor him by giving his name to a proposed railroad junction down in Fresno County on a line then being surveyed. The railroad was the Fresno-Porterville branch of the Southern Pacific which was completed in July, 1888.

The SP erected a depot at Sanger Junction and offered town lots for sale. The development achieved only moderate success until the news came about the flume. The effect was like that of a gold strike! Sanger became a boom town overnight. Buildings sprang up at a rapid pace and labor was in short supply. A town lot that had been worth only $300 a few months before was sold for $4000! In 1890, more than 75 structures were erected within the town limits at a cost exceeding $125,000.

Joseph Sanger never saw his namesake. He returned to Indiana following the convention and died in 1899.

* * *

Once the flume route was settled, Smith turned all his efforts toward getting the immense project completed. He was hopeful of finishing the high altitude sections before winter arrived, and work was pushed in that regard during the late summer and fall of 1889. Most of the lumber used in building the flume was floated board by board from the mill to the construction point. Work

This badly faded but extremely rare old print shows flume construction in 1889. Building the flume through the rugged Kings River Canyon was a remarkable accomplishment. (Fresno County Historical Society)

Early tourists at Smith and Moore reservoir situated about a mile above the upper mill.

crews camped at the job after working from dawn to dusk.

By September 5, 1889, five miles of flume were completed. By the end of November, the 10-mile mark was passed. Heavy rain and snow began to slow the progress. Builders worked with numbed fingers as chilling water poured from the flume and covered materials and tools alike with a layer of ice. Keeping the shivering workers on the job was difficult. About 30 construction men were employed along with 18 Chinese graders.

"I never want to build another flume," Hiram Smith complained. "There is too much worry and trouble with it, especially in winter. Why, we have not been able to work more than a third of the time and when the weather did afford us an opportunity to do so, the lumber and everything else was so covered with ice and snow that we made poor headway."

By the first of April, 1890, three months behind the anticipated schedule, the flume had almost reached the Kings River crossing where easier going would be encountered. The end was in sight.

In the meantime, contracts were let for the reservoir at Mill Flat Meadow. A concrete and stone dam was started in August, 1889, with the idea being to hold back the waters of Mill Flat Creek thus making an artificial lake about 20 feet deep. The purpose was to provide a controlled supply of flume water but there were hopes of capitalizing on the area as a summer resort.

"Our lake will be plentifully stocked with fish," a company spokesman boasted. "New hotels will be built next summer and boats provided to make the resort first class in every respect, especially if we get a new wagon road."

The reservoir was situated at a 5300-foot elevation and it was bitter cold during the winter months. Heavy snows prevented construction at times.

"The reservoir contractor has trouble getting white men to work," the Visalia Weekly *Delta* observed late in 1889. "There are now 75 Chinamen employed. A visitor to the site tells us that he never saw such a gang of slow pokes in his life. A crew of 20 white men could do more with little effort."

One phase of the complicated operation was proceeding smoothly for Smith and Moore and that was the sawmills themselves. Teams were busy hauling logs from the timber and several chutes were in use. The company was its own best customer as the flume required nearly all the lumber that could be cut.

"It is most pleasant in the mountains this summer," the Fresno *Expositor* reported in 1889. "There are now about 25 houses near the upper mill and the area resembles a small town. Several dozen campers are in attendance."

A favorite pastime was riding the flume from the upper mill to the lower mill, a distance of two miles. The *Delta* correspondent described the activity: "A box 16 feet long, V-shaped, holds two or three passengers for flume riding. In some places grades are quite steep but even ladies ride the flume. It takes 10 minutes and is most exciting!"

On October 3, 1889, a "thrilling race" was held between the two mills to see which could produce the most lumber in a single day. The upper mill cut 76,000 board feet against the lower mill's total of 72,000 board feet.

"Our operation exceeds all expectations," Smith proudly reported. "We are still hopeful of achieving year-round production at the lower mill."

The upper mill, sometimes called the Sequoia Mill, ran from July 6 to November 18, 111 days

High trestle under construction on the great Sanger flume in 1889. Workers braved freezing weather, rattlesnakes, poison oak and rocky terrain in order to complete the 54-mile route. (Fresno County Historical Society)

of work; the lower mill, also known as the Abbott Mill, worked from August 26 until December 2, a total of 67 days. The average cut was 45,000 board feet per day, per mill, for a grand total of about eight million board feet during the first season. Nearly all the production was used in building the flume. Despite Smith's avowed intentions, no winter operations were conducted.

(There has been some confusion regarding the exact location of the lower mill. It was situated on Abbott Creek, a name which was incorrectly applied to a nearby stream on most early maps. This error has been rectified on modern maps.)

* * *

Once the winter snows left the Sierra in the spring of 1890, all phases of the gigantic enterprise began to tie together.

Construction of the Sanger facilities began in April, 1890, and comprised a box factory, planing mill, engine house and drying yard. These were finished in July when the flume itself was only 15 miles from its destination. The reservoir

Flume as it paralleled the Kings River, part-way down toward the valley. Feeder troughs kept the flume replenished with water along the way. (Ross Phillips Collection)

high in the mountains picturesquely called "Sequoia Lake," was brimming with water and ready for use.

"The best system for conveying lumber that inventive genius can devise is nearly ready," the Sanger *Herald* exulted. "Smith and Moore will soon make us the industrial center of the county. They have unlimited financial backing for this wonderful project."

* * *

On September 3, 1890, just 13 months and six days from the time the first timber was put in place, the flume was at last completed. A great celebration was held at Sanger.

"Come and ride the big flume," invited the posters and advertisements that told of the dedication ceremony.

"Visit our Flumeopolis," urged the *Herald*.

Visitors poured into Sanger from all over the valley. A special three-car train arrived from Fresno jammed with excursionists and politicians.

By noon on dedication day, a crowd of "several thousand" was in attendance. Flowery speeches were made by Senator Goucher, Judge Northrup and other dignitaries. The Expositor Military Band presented a concert. Festivities included a fight held by two middleweight boxers imported from Los Angeles, and an evening dance in an open-air pavilion which was hastily constructed for the occasion.

During the activities, the celebrants consumed 10 sheep, three bullocks and 300 loaves of bread. All saloons reported capacity business.

It was a glorious day.

FLUMEOPOLIS

EXCURSION

BARBECUE AT SANGER!

SPECIAL TRAIN

Will Leave FRESNO at 9:30 A. M. September 3d,

Returning, Leave Sanger at 7 P. M.

FARE: Round Trip, - - $1.00

COME AND RIDE IN THE BIG FLUME!

Dancing, Barbecue, Speaking, Flume, Music, Refreshments.

Special train from Fresno was run by the Southern Pacific on the occasion of the completion of the flume. The above advertisement from the September 2, 1890, edition of the Fresno **Expositor** illustrated the interest generated by the flume project. (Spencer Crump Collection)

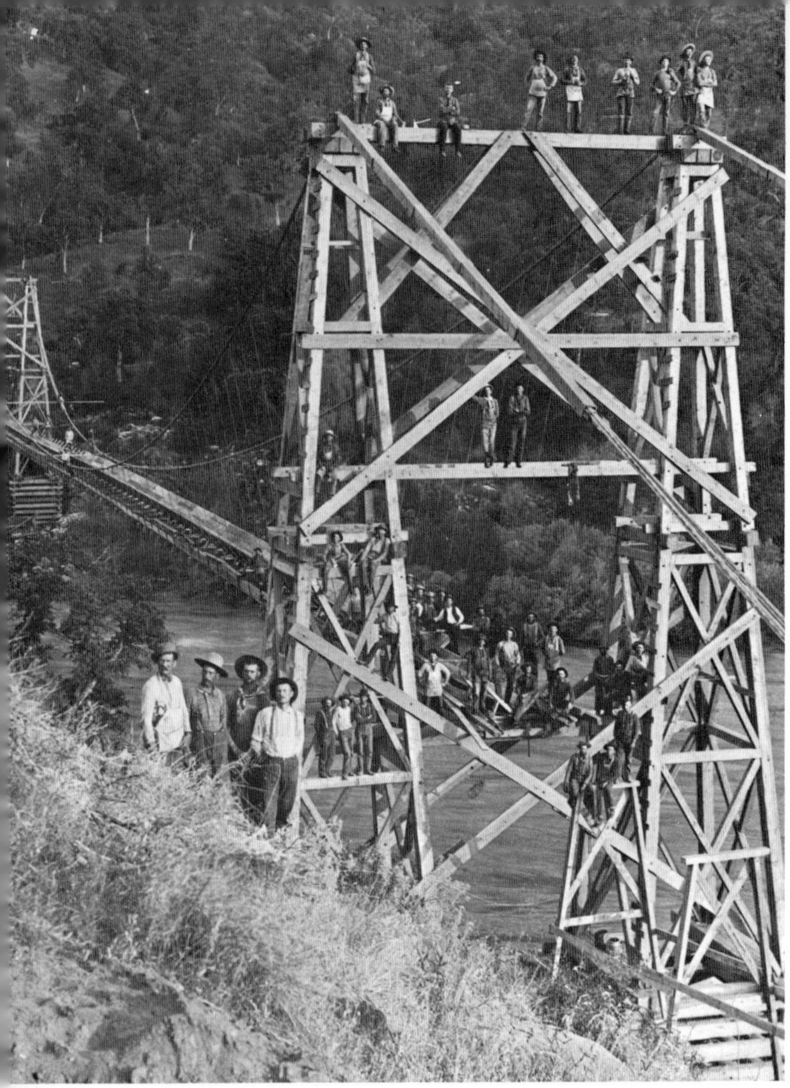

The flume bridge across the Kings River, built by the Pacific Bridge Company of San Francisco, was completed in June, 1890. It measured 451 feet between towers with seven 1⅜-inch steel cables on each side. Originally constructed without a stiffening truss (above), the bridge proved unsatisfactory because water would not run evenly but instead caused a series of waves which resulted in extensive oscillation and spillage (below).

The flume went into regular service on September 24, 1890. During November, 2,255,000 board feet of lumber floated down from the mountains to Sanger. In January, 1891, 2,762,000 board feet were sent down the flume. Most of the 1890 output of the two mills had been stored in the yards awaiting completion of the flume.

"We are well satisfied," bragged a Kings River Lumber representative. "From September 24, 1890, until February 4, 1891, we have shipped nearly 12,000,000 board feet to Sanger. This is the best showing of any flume in the west, and we have had few washouts."

The Sanger flume was indeed a tremendous accomplishment. Its total length of 54 miles not including feeders or switches made it one of the longest in the world.

* * *

Early flumes were used for several purposes: some carried water from rushing mountain streams for irrigation or power projects; others transported logs from remote forest areas to sawmills in the lowlands. Lumbermen eventually determined that by locating the mill itself in the timber, all waste material could be left at the mill site. By shipping only usable lumber out of the mountains, transportation costs were drastically reduced.

This application resulted in the building of tremendous lumber flumes. The longest and most costly were in California.

Flumes were originally box shaped with square sides and bottoms. These were excellent for carrying logs and water. Lumber, however, tended to jam in this type of framework. A Nevada logger pioneered a V-shaped flume which proved highly successful. The V-flume was almost self-clearing as the backhead lifted the planks into the widening top of the flume. Additionally, it took less lumber to construct, and more importantly, less water to operate than a square-box affair.

California lumbermen copied the Nevada idea. Small flumes were built with some success near the Sacramento River in the early 1870's. In the southern Sierra, the California Lumber Company, forerunner of the Madera Flume and Trading Company, began its nearly 60-mile flume in 1874.

A stiffening truss, also built by the Pacific Bridge Company, remedied the difficulty first experienced at the Kings River crossing. Photo at right shows completion of this project. No further trouble was reported with the bridge even under heavy loads (Fresno County Historical Society).

Crew clamping lumber at upper mill shipping yard. Note clamp on bundle at right side of picture. (Kenneth Rea Collection)

There was thus some precedent to guide Smith and Moore when they first considered the problem of bringing their lumber to market from the mountain slopes. Nevertheless, their project was daring and very expensive.

The Sanger flume required 9,000,000 board feet of the finest knot-free lumber to build and cost a reported $300,000. Clear lumber was a necessity as any blemish would allow water to leak out.

Like other V-flumes it was put together in 16-foot sections called "boxes," and the desired grade was one inch of fall per box or 27 feet to the mile. Because of the uneven terrain, this grade was seldom maintained despite the use of trestles nearly 100 feet high in places. Sometimes the fall was so abrupt that the lumber was swept along at a speed of 50 miles an hour. In the lowlands near Sanger, the ground was so nearly level that a man could easily walk faster than boards were carried in the flume.

Flume House somewhere along the 54-mile route. Trough is visible at left. (Mac Grosse Collection)

The size of the trough varied according to the steepness of the slope. From the upper mill to Station Three, the flume was 32 inches on each side, lined with two 1½ x 16 inch boards. From Station Three to Station 10, the size was 42 inches on each side. This increased to 48 inches from Station 10 into Sanger. On this portion, the top of the trough was more than five feet across and the water was nearly three feet deep. A telephone line stretched the length of the flume and station numbers were used to designate phone outlets. There were 11 stations in all.

The upkeep of the flume was handled by "herders" who patrolled "beats" of about three miles each. These herders, living in isolated shacks at vital spots along the route, were responsible for separating lumber, preventing jams and repairing leaks. As they ran along narrow catwalks only six to eight inches wide, they kept the lumber moving with the use of picaroons — special hooked tools fashioned in the blacksmith shop for the purpose. The catwalks had no railings and there was nothing to prevent a fall of 100 feet or more in places save the agility of the herder.

It was a lonely job. Often there was no contact with the outside world for weeks except by telephone. About 40 herders were employed.

Lumber was placed in the flume at the upper mill. It was first clamped into bundles a foot square using 12-inch, pointed iron clamps. Wooden wedges were then driven into the bundles to tighten the clamps and form a solid package. Lengths were normally 18 to 28 feet. Boards were cut slightly oversize so the battered ends could be trimmed at the Sanger finishing mill.

As the bundles were launched into the slowly running flume, they were tied together with manila rope, end to end, into trains of from three to six bundles. The long journey to Sanger began. At the lower mill two miles below, additional bundles were hooked to the train. Farther down, flume herders held up the bundles at slack points along the line until subsequent trains arrived. These larger groups were then joined together until the lumber finally moved into the valley in an almost continuous train. At Sanger, the flume was unloaded and the water turned into a branch of the Fowler Switch Canal for irrigation use under an arrangement with the owners of the riparian rights.

The idea of clamping the lumber into bundles was developed by William Thurman, a founder of the California Lumber Company at nearby

Excellent C. C. Curtis photo shows upper mill and drying yards. Flumes originated at various points in the yards and joined into a single route just beyond the shipping area. (Harold G. Schutt Collection)

Madera. Fluming loose boards had proved impractical in that early operation and the clamp arrangement was the result. Thurman, in fact, patented his clamp but whether he was able to enforce royalty payments is unknown.

Boards too wide for fluming and special high-grade pieces were hauled to Sanger by freight wagon. Teams pulling heavy loads made their precarious way down the primitive road from the mill to the valley. On their trip back to the mountains, the wagons returned the flume clamps and wedges to be used again.

* * *

As Smith and Moore made ready to open the 1891 logging season — the first year of full operations — they must surely have had some misgivings about the great flume that had been so laboriously erected along the Sierra slopes. The slender wooden trough that snaked its way down the mountains, across huge ravines, over rivers and streams, and through dense forests was extremely vulnerable to attack by flood, fire and snow. It represented a fortune, yet its value was still not really proven.

Outwardly, at least, the investors appeared confident:

"The flume and mills are as nearly perfect as human ingenuity and foresight can make them. They are adding daily to the material wealth of the county and the state. A spirit of enterprise is everywhere."

The Kings River Lumber Company was ready, and the denuding of the Sierra could soon begin in earnest. The ring of the woodman's axe and the echoing cry of "Timber-r-r-r" awaited only the coming of spring.

Flume herders with their picaroons at some point in the mountains. Apparently the flume is being repaired in this picture. (Dick Challacome Collection)

CHAPTER THREE

RAILROAD TO THE REDWOODS

SANGER, SELF-STYLED "Flumeopolis of the West," teemed with activity in the spring of 1891. Groups of rugged lumberjacks dressed in rough work clothes passed through the town almost daily on their way to the mountain mills of the Kings River Lumber Company 50 miles to the east. New settlers swelled the population of more than 2000 with each incoming train. Business was thriving and the future of the community appeared full of promise.

"We are the healthiest three-year-old in the state," proclaimed the Sanger *Herald* with understandable pride. "Besides the great lumber mills, we have two excellent hotels, four splendid brick business blocks, several large grain warehouses, a bank and a first-class weekly newspaper. Our raisin and fruit industry will soon be second to none in the county and our general growth is unprecedented."

But the coming of urban civilization to the southern San Joaquin Valley did not immediately alter the frontier habits of the residents. A goodly portion of the male population still carried revolvers — not openly on the hip as before, but stuck in a waistband or hidden in a coat pocket. Brothels, opium dens and dance halls operated without interference. Gambling was widespread and nearly every saloon provided games of chance.

Stagecoach holdups, train robberies and other outlawry spiced the news on frequent occasions. San Franciscans were so eager for information about the "wild west" that several metropolitan newspapers sent correspondents to cover the more exciting occurrences first hand.

In 1891, members of the well-known Dalton gang of Oklahoma were arrested in Fresno and charged with complicity in the robberies of three Southern Pacific trains near Tulare. After a lengthy trial, Grat Dalton was found guilty. While awaiting sentence, he and two companions sawed through a cell bar under cover of some loud group-singing and replaced the bar with a length of blackened broomstick. At a propitious moment the next night, the outlaws removed the substitute barrier and escaped into the darkness. Despite

Stagecoach on the Sanger-upper mill road. There were few turnouts on the one-lane route and the trip was very dusty. (Mac Grosse Collection)

heavy pursuit, Grat made it safely out of the state.

The Daltons, cousins of the Younger brothers, another celebrated group of western desperadoes, have received more notoriety from television and movie "biographies" than their short, 18-month careers of crime seem to deserve. But the gang, headed by Grat, Bob, Bill and Emmett Dalton was indeed a formidable one and participated in a number of holdups and killings in Oklahoma before being wiped out at Coffeyville, Kansas, in 1892. In this affair, the Daltons attempted to simultaneously rob two banks during daylight hours. Bob, Bill and Grat were shot dead by an aroused local citizenry while Emmett was badly wounded. After serving a lengthy prison term, Emmett took to the lecture platform to warn against a life of crime. He died of natural causes in 1937.

The infamous Visalia bandits, Chris Evans and George and John Sontag, were at the peak of their depredations during the early 1890's and often hid out along the flume of the Kings River Lumber Company when the sheriff got too close for comfort. Because of alleged mistreatment by the Southern Pacific, the fugitives took particular delight in harassing the railroad at every opportunity. Between "jobs," they were frequently seen at the mountain mill camps as they had many sympathizers in the company's employ.

In 1892, Evans and George Sontag held up the stage from Sanger to the upper mill. It was said that the bandits often stopped stagecoaches to "see if there were any lawmen aboard that we can kill." For almost a year, the activities of the outlaws made headline news throughout much of the state.

Evans and the Sontag brothers were put out of action in 1893 by determined lawmen (John Sontag was shot to death; George Sontag and Evans were imprisoned), but lesser-known desperadoes continued to plague valley residents until the turn of the century. Gambling, prostitution and other illicit activities persisted for another decade on an ever-diminishing basis.

* * *

By the summer of 1891, the Kings River Lumber Company was in full-scale production at both the Sanger yards and the two mountain mills. At Sanger, more than 200 men were employed in the box factory, sash and door factory, planing mill, engine house and drying yards. The facilities covered 67 acres. As the raw lumber arrived from the Sierra in the great flume, it was lifted out manually by "dumpers" and sent down slanting skidways to the main platform. Running along the 14-foot high platform was a tramway on which the lumber was moved by small cars to the different parts of the yards. A mile of tramways intersected the yards and on either side were huge piles of yellow pine, mountain pine, sugar pine, redwood and cedar, often 30 feet or more in height.

Lifting the heavy, wet bundles of lumber from the flume was a back-breaking job that required strong and sturdy men. The normal work schedule was from 6 a.m. until 6 p.m. with an hour off for lunch, six days a week. There were no unions or "fringe benefits" and a day's pay of $1.50 was well earned.

The lumber was segregated by grade upon arrival. The knotted, poor boards went directly to the box factory while the better class of lumber was sent to the planing mill for finishing. Power for the various machines was supplied from two engines having a combined output of 260 horsepower driving a fly wheel 14 feet in diameter. A band running over the 40-inch surface of the fly wheel turned the main shaft to which all equipment was connected. Sawdust from the factories provided a major portion of the fuel and was driven by fans through an underground pipe into the furnace vault.

Once the finished lumber was sufficiently dry, it was sent out by railroad car over the Porterville branch of the Southern Pacific to markets along

ABOVE: Bundles of clamped lumber shooting down the flume. (Jean Hume Browning Collection). BELOW: Flume terminus at Sanger where "dumpers" unloaded the bundles for stacking. (Bill Reid Collection)

the coast and overseas. About two miles of Southern Pacific switch lines intersected the drying yards between the tramways so that lumber could be loaded at convenient points. During the early years of the company, many shipments were made to Australian customers. Sailing vessels were chartered out of San Francisco harbor.

"Our payroll exceeds $12,000 a month, and we have the most efficient plant in the west," Hiram Smith announced in 1891. "Many labor-saving devices have been put into use and we are ready and able to meet all competition."

* * *

High up in the Sierra Nevada, a force of 300 men was busily engaged in felling timber, hauling logs to the mills, and preparing rough lumber for the 15-hour flume ride down to Sanger. In order to reach forest lands farther from the mills, two railroads were built into the woods in 1891. From the upper mill south along Mill Flat Creek, a crude tramway made of wooden rails covered with strap-iron, ran to a log landing about a mile away. Empty cars were hauled by means of horse teams up to the landing where they were loaded with logs supplied from chutes that probed the surrounding timber. The return trip to the sawmill was accomplished by gravity since it was all downhill.

At the lower mill, a steam-powered, narrow-gauge railroad connected with a little incline and a landing two miles to the west across Mill Flat Creek. Power on this line was furnished by a 36-ton, Shay locomotive that was purchased new from the Lima Locomotive and Machine Works in Ohio. The Shay was partially dismantled upon arrival in the valley, and tediously hauled piece-meal by team and wagon to the Sierra location. With it came a quantity of disconnected railroad log buggies equipped with hand brakes and link-and-pin couplers. These cars were also used on the horse-powered tramway.

Shays were geared logging engines that had only recently been perfected by a Michigan logger named Ephraim Shay especially for use on rail-roads where rough trackage and steep grades were a problem. Their speed was low but their pulling power was enormous. The locomotive was called

The horse tramway that ran from the upper mill toward Sequoia Lake. Note chute and teams at left which brought the logs to the landing. "Horse-power" railroad shown at right. (Harold G. Schutt Collection)

ABOVE: Railroad that operated in 1891 from the lower mill to woods landings. Oxen in backbround brought logs to the train route. (Annand Collection)

BELOW: Sanger yard showing the drying area with mill in distance. Plant and yard covered 67 acres. (Jean Hume Browning Collection)

the *Sequoia* and the two-mile rail line was grandiosely named the Sequoia Railroad.

Logs were initially brought by chute to the various landings served by the railroads using oxen or horse teams, but donkey steam engines soon replaced the animals at all woods camps. These vertical-boiler machines powered steel cables for a considerable distance and enabled the loggers to reach otherwise inaccessible forest tracts with ease. During the first two seasons, very little redwood was cut and the majority of timber was comprised of fir, pine and cedar.

In 1891, the Kings River Lumber Company produced nearly 20,000,000 board feet of lumber but Smith and Moore were not satisfied with the results.

"We must increase our volume to justify the enormous investment and expense of the mills, railroad and flume," a company official observed in December, 1891. "We plan to move our railroad to the northeast where the really big trees are located. Next season should be highly profitable."

In April, 1892, the Visalia Weekly *Delta* reported that 50 Chinamen and 20 white men were building a "seven-mile railroad" running northeast from the upper mill. During the grading, an unfortunate accident occurred. One of the Chinese workmen placed three charges of black powder in an obstructing rock. Only two charges exploded. After a short wait, the Chinaman indiscreetly approached the third charge to investigate the problem. As he reached the area, the last charge suddenly went off "blowing the heathen to bits." It was intended at first to raft the remains down the flume to Sanger but for some reason they were transported by wagon to Visalia instead.

The new railroad was in use for the 1893 season and logs were brought to landings along the route even as the tracks were being pushed farther to the northeast toward the eventual terminus — a high ridge about five miles away on the very edge of the biggest redwoods of all. There were 23 trestles and innumerable curves on the little rail line that wound through the virgin forests. Construction was of the most rudimentary sort and very light, 30-pound rail was used. The single Shay locomotive was in constant use as it made its way back and forth carrying log loads to the mill and empty cars back to the timber.

In July, 1893, an incline was built from the end of the railroad up a steep grade to the top

ABOVE: Interesting rear-view photo of the "Sequoia" Shay locomotive. This engine was equipped with link-and-pin couplers and burned cordwood. BELOW: Building the original railroad grade the hard way — with shovels and wheelbarrows! (Bill Reid Collection)

Load of logs (some redwoods) near the end of the Sequoia Railroad. Mountainside chute in background appears very steep. That's a pretty but unusual conductor on the cab steps! (Harold G. Schutt Collection)

of the 6700-foot mountain which is now called Hoist Ridge. The idea was to pull timber out of the valleys adjoining the ridge and lower it down the incline to the waiting railcars below. Several of the largest donkey steam engines were used at the summit for this purpose. Although some redwoods were being cut in other areas, the hoist arrangement marked the first extensive redwood logging for the company. Since there was a limit to what the donkeys and the railcars could handle, logs more than eight feet in diameter had to be blasted into sections with black powder before they could be moved. Holes were first bored into the downed trunks using long augers, and powder then inserted at proper intervals. There was a great amount of loss because of splitting in this arrangement, but no one seemed to worry about waste as long as the mills were kept busy.

* * *

A summer resort of considerable size was gradually growing up around the upper mill. The crisp mountain air and beautiful scenery proved very attractive to residents of the hot central valley. It was not an easy journey by either stage or buggy, but hundreds of hardy vacationers from Visalia, Fresno, Sanger and other nearby towns made their way into the cool Sierra during the summer months. Some camped at Sequoia Lake but most settled at Millwood, which was the name given to the new community.

There was much for the visitor to see and do. The two mills were in full production and the whir

Blasting a Big Stick

To handle the redwoods that were too big for either the railcars or the mill to handle, it was necessary to blast the logs into smaller sections. Photo below shows loggers drilling holes into a big log preparatory to packing with blasting powder. Augers hanging at end give an idea of the depth of the holes. (Harold G. Schutt Collection). Picture at left was taken at the exact moment of a blast. Hopefully the log would split into relatively even sections. (Bill Reid Collection)

Early day belle out for a stroll along the flume catwalk. There was nothing to support the walker as the flume sides were too low to use as a railing. (Jean Hume Browning Collection)

of the big saws could be heard up and down Mill Flat Canyon. Every morning a company employee released water into the flume from a small pond which had been built to provide an auxiliary reservoir close to the mill. At 6 a.m., workers began loading bundles of clamped lumber into the slowly running trough for the long trip down the mountains. Shipping continued until late afternoon. The final bundle of the day was flagged with an upright board called a "joker." As the "joker" passed each flume station, it was reasonable proof that the route was still intact and unblocked. Herders reported its progress by telephone. The flume carried 250 inches of water per second and was replenished by four feeder flumes along the line. Daily capacity was 250,000 board feet but shipments seldom exceeded 140,000 board feet.

Millwood boasted a shoemaker, photographer, barber, blacksmith, doctor and an undertaker who advertised that he would accept cases "day or night." There was a post office, general store, butcher shop, livery stable, several saloons and later, two hotels. The first and larger hotel, the Sequoia (everything seemed to be named "Sequoia"), stood slightly south of town and above the road up to Sequoia Lake. A Visalia man converted a large company building into a rival hotel called the "Red House" and the Sequoia soon followed suit and became known as the "White House."

There was a summer school with about 40 pupils, a slaughter pen to supply fresh meat, and a "red light" district with its own little village about a mile to the south. It has been said that a trollop or two could also be obtained at the hotels, although the name "Red House" supposedly signified the color of the exterior paint rather than the internal activities.

Several hundred lumber company employees lived in the community. The married men and their families had individual shacks while the single workers resided in rough dormitories and ate at a common cookhouse. There was plenty of gambling, drinking and fighting among the lumberjacks who were a rough and tumble group. It was "standing room only" at the saloons and brothels of a Saturday night, but the men were still ready for a hard game of baseball on Sunday, the only day off. The more genteel amusements included boat rides on Sequoia Lake and croquet contests on the community court.

In 1894, a new road was built from Sanger to Millwood by Fresno County, and this considerably eased the arduous mountain trip. The arrival of the stage with its mail and passengers was always an exciting event in the isolated Sierra.

It was a gay life at Millwood in the 1890's and so appealing to the city dwellers that the summer population reached 1000 people or more during peak times. The Merced *Sun* reported in 1892 that "railroad people were looking into the feasibility of building a railroad from the San Joaquin Valley to Millwood to capitalize on the pleasant and invigorating clime of that popular resort." This optimistic scheme never got beyond the planning stage but it illustrates the appeal that the Sierra exerted at a very early date.

* * *

Things were not quite so gay for the Messrs. Smith and Moore in 1893, however. They had

ABOVE: The Red House Hotel at Millwood. Sign advertises beds and meals for 25¢ each. (Bill Reid Collection)

BELOW: Big Trees Stage arriving at the Sequoia Hotel (White House) on a summer afternoon.

ABOVE: Redwoods atop Hoist Ridge ready for lowering to railroad and then to the upper mill. This was forerunner of the big Converse Hoist operation.

BELOW: Overall view of the flume terminus at Sanger. Drying yards occupy space on both sides of flume. (Dick Challacombe Collection)

unwittingly completed their complicated and expensive enterprise just as the country entered upon one of its most severe periods of depression. Business began to dip in 1892 and a year later, the panic struck full force. Many banks closed their doors; businesses failed in large numbers, and thousands of farms and homes were foreclosed. The construction industry was at a standstill and there was no demand for lumber at any price.

Although Smith and Moore owned all the stock of the Kings River Lumber Company, they were faced with heavy payments on the mortgage as well as a considerable monthly overhead. In an effort to procure immediate cash, electric lights were installed at the mills early in 1894 so a large Australian order could be quickly processed using two shifts. But the overseas shipments were not enough to make up for the absence of local business and by the summer of 1894, the company was in serious financial difficulty.

"Great dissatisfaction now prevails with the mill hands of the Kings River Lumber Company," the Visalia Weekly *Delta* warned in June, 1894. "If some satisfactory arrangements for back wages are not made at once there is every possibility of serious trouble arising."

On July 5, 1894, valley newspapers reported that the Sequoia Mills were closed down under attachment and that it appeared as if the company might be permanently out of business. About 600 employees were laid off — some with back pay owing.

"Your reports regarding the Sequoia Mills do us great injustice," Hiram Smith angrily replied in the Fresno *Expositor* of July 12, 1894. "Both mills are running night and day and will continue the month out. There is no attachment and the men discharged have so far all been paid."

But Smith's protestations to the contrary, the Kings River Lumber Company's plants were bolted up a few days later under court order. Austin D. Moore hastily set to work in San Francisco in an attempt to rescue his tottering financial empire. With the help of loans from friends, business associates and banks, a reorganization took place that culminated in the establishment of a successor to the destitute Kings River Lumber Company. On October 15, 1894, articles of incorporation were filed in San Francisco under the name of The Sanger Lumber Company, capital stock $600,000, with the following directors: A. D. Moore, Walter Young, Ivory F. Littlefield, H. C. Smith and Charles R. Johnson. Johnson was the founder and president of the Union Lumber Company, owners of a large mill and railroad at Fort Bragg, California, which is still in existence. In recent years, its picturesque little railroad, affectionately called the "Skunk" because of the gasoline railbusses which carry tourists through the coastal redwoods, has achieved nationwide recognition.

Smith and Moore were heavily obligated to their backers in the new arrangement. They had pledged great quantities of Sanger Lumber Company stock as collateral for the promissory notes which enabled them to reopen. Included in the complicated arrangement were mortgages and loans involving other properties owned by the two men such as the Port Discovery Mill in Washington.

The nation's economic outlook brightened a little late in 1894 which is perhaps one reason why Moore was able to obtain fresh backing, but the situation took a turn for the worse in the summer of 1895. There was another season of heavy losses which again placed the two San Francisco lumbermen in the position of being unable to meet their obligations. This time there was no way out. The holders of the various notes and mortgages exercised their rights and took over The Sanger Lumber Company in August, 1895.

Baseball was a popular summer pastime for the loggers. The White House Hotel nine is pictured below. (Harold G. Schutt Collection)

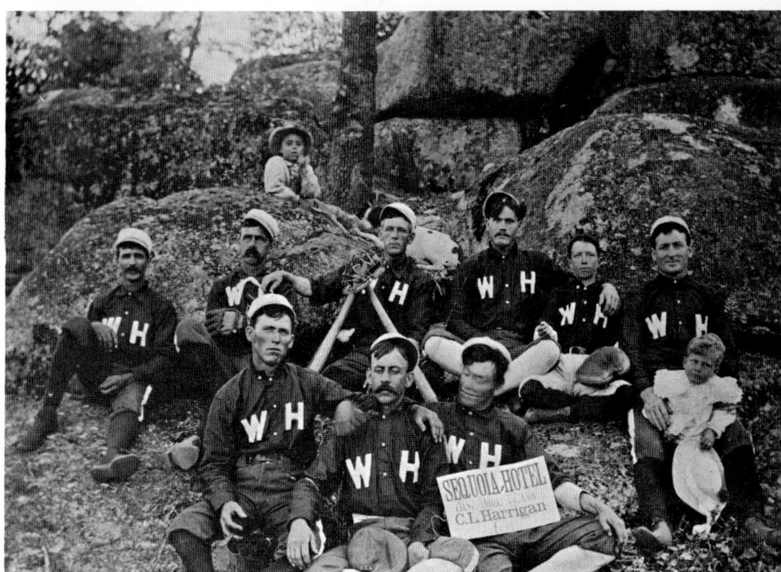

The lower mill in 1891. Railroad at left brought logs from outlying landings to mill. This mill operated from 1890 through 1896 and was abandoned upon the move to Converse Basin. The railroad was shifted to the upper mill in 1892. (Paul Spivey Collection)

Moore was finished by this action and retired, a broken man. Smith, in an attempt to salvage something from his investment, agreed to stay on as manager for the creditors until it was determined what direction the lumber business would take.

In 1896, a cut of 22,000,000 board feet was reported which was not sufficient to meet expenses. It was decided that the only way to realize anything on the immense investment was by placing a sawmill in the very heart of the biggest redwoods and making an attempt to log the huge trees. Some of the giants near the end of the railroad, just over Hoist Ridge, contained nearly a half-million board feet of lumber in a single tree!

With the permission of the creditors, Smith kept a crew of 100 men working all winter on the project. Improvements were made to the huge, permanent hoist which had replaced the steam donkeys atop Hoist Ridge two years previously in anticipation of the eventual move. The incline railway was continued down the back side of the ridge into a natural bowl called Converse Basin. Here were 5000 acres of some of the largest trees in the world! The new sawmill was located near the center of the basin straddling Converse Creek, about the only flat place in the area.

"We are making excellent progress with our new mill," Manager Smith announced in February, 1897. "This season should see the start of the biggest and best logging operation in the whole Sierra region."

If nothing else, Smith was always the height of optimism in his public pronouncements. The creditors hoped that this time he would be right.

ABOVE: View of Millwood drying yard. Railroad tracks at center right. The town buildings are beyond the right edge of this picture.

BELOW: Almost the same scene during winter. Heavy snows blanketed the logging area during the off season. (Both Dick Challacome Collection)

THE FOURTH OF JULY

The biggest celebration held during the logging season was on the Fourth of July. On most such holidays, a picnic train carried the loggers and their families to some remote woods location. Occasionally, the happy celebrators extended their festivities into the subsequent work day to the disapproval of the management. A typical excursion train is shown at the left on a Fourth of July about the turn of the century. Below is pictured a load of big redwood logs on July 4, 1893. Note the markings on the log ends as to the total number of board feet in each log. The train is probably on its way back from Hoist Ridge. Next Page (top) is photo of another fourth celebration a few years later taken on the railroad between Millwood and the end of the line. Sequoia Hotel is gaily bedecked in holiday bunting in bottom picture. (All photos Harold G. Schutt Collection except upper right which is Sabina Olsen Collection)

☆

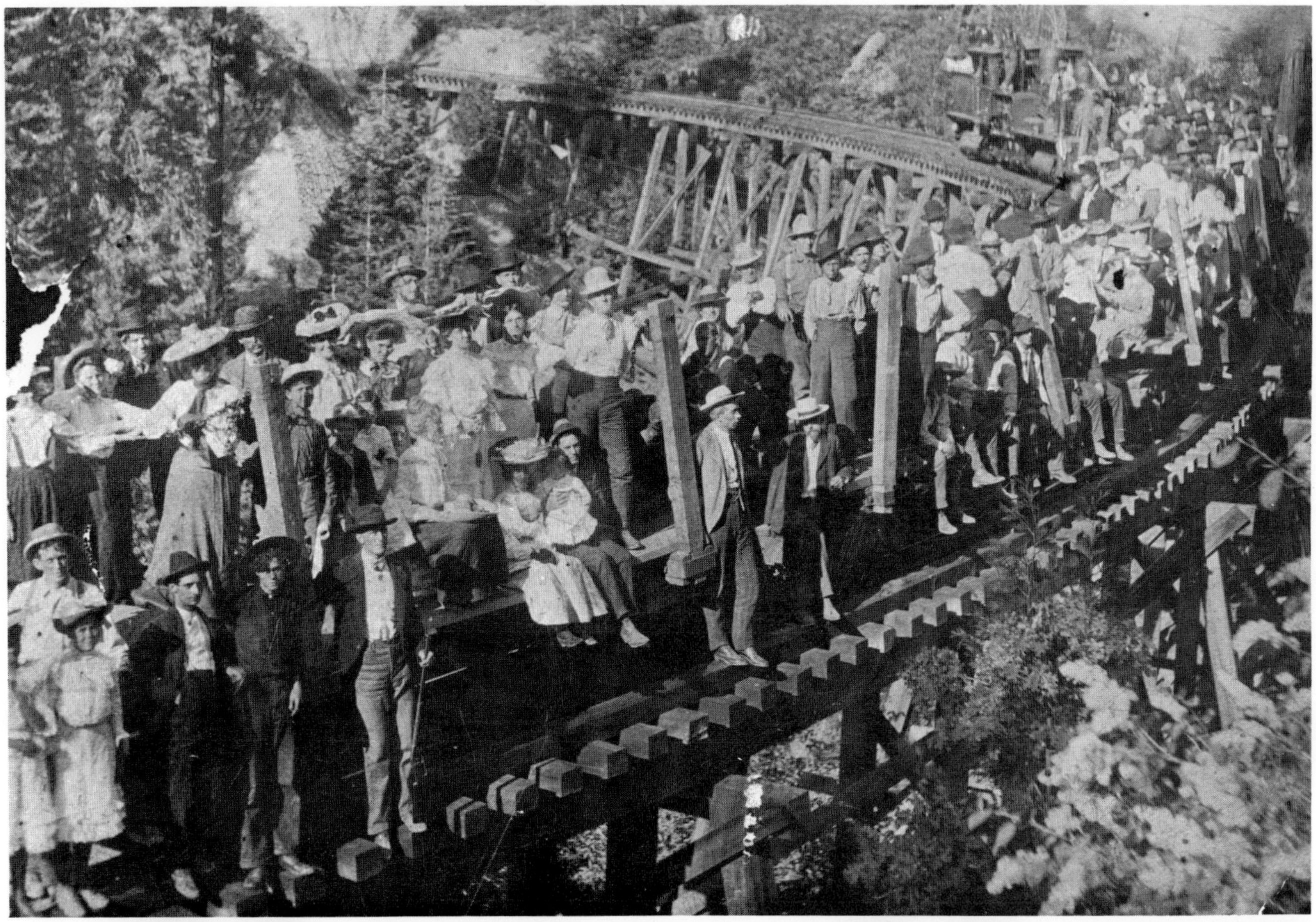

The "Sequoia" at a long landing between Millwood and Hoist Ridge. This landing was known as Andy Berry's Landing, so-called because the Berry Brothers here ran a logging show under contract with The Sanger Lumber Company. Timber was carried by railroad to Millwood for cutting. (Harold G. Schutt Collection)

CHAPTER FOUR

THE BASIN CALLED CONVERSE

CHARLES P. CONVERSE was a character right out of the pages of a wild-west novel. A man of striking appearance — six feet tall, more than 200 pounds and possessed of great physical strength — he seemed to stir up controversy wherever he went.

Converse was beset by ambition and jumped quickly from one venture to another in his quest for success. Although rather sloppy in attire and with generous features topped by a great shock of unruly hair, he could nevertheless be a glib and persuasive talker when the mood struck him.

Young Converse migrated to California from his native Michigan during the 1849 gold rush. In 1851, he opened a general store at Coarsegold selling supplies to the local miners. A year later, he was running a ferry on the San Joaquin River near present Friant. This was soon followed by a logging operation at Crane Valley (now Bass Lake), and a period of farming near Millerton.

Converse was continually on the edge of trouble. There were those who felt he was responsible for the murders of two prosperous miners along the San Joaquin River in the early 1850's; he was fined $250 for stuffing ballot boxes in the general election of 1861, and there were several arrests at various times for illegal gambling activities.

Despite his unsavory reputation, Converse was awarded a $17,000 contract in 1866 as low bidder in the construction of a two-story courthouse with an "escape proof" jail at Millerton, the original seat of Fresno County. A jail erected there a decade earlier was so poorly built that prisoners could not be left unattended.

Soon after the new courthouse was completed, Converse became heavily involved in the local sheriff's election between James Walker and the incumbent, J. Scott Ashman. Partisan feeling was high and backers of both candidates were openly armed. Converse was a vehement Walker supporter and not the least bit shy about expressing his preference at every opportunity. An outbreak of violence seemed imminent throughout the day of election. When the polls closed at 3 p.m., Converse repaired to Payne's saloon to await the results. Suddenly, from outside the building, a heavy cobblestone was thrown at him, missing by the narrowest of margins. Converse instantly drew his revolver and fired at his fleeing assailant but also missed. As he was thus engaged, William Crowe, a local barkeep and ardent Ashman adherent, struck Converse a powerful blow on the back of his head with a "slung shot." It was later testified that only Converse's tremendous strength and vitality and his thick felt hat saved him from the assassin's attempt on his life. Though knocked to his knees and badly dazed, he doggedly arose and fired at the running Crowe, dropping him dead in the street with a well-aimed shot. He whirled and fired again, knocking the hat from the head of the rock thrower, one John Dwyer, who managed to escape behind the corner of a nearby hotel. Converse braced the gathering throng seeking further attackers, but was quickly disarmed by men in the crowd. Ironically, he became the first prisoner to be incarcerated in his own "escape-proof" jail!

It was some satisfaction to Converse that his man Walker won the election but he was extremely bitter about his arrest because of the circumstances surrounding the affair. He was indicted for murder on August 10, 1867, but the charge was subsequently lessened to assault with a deadly weapon. On December 17, after a short trial, the jury found Converse "not guilty" by reason of self-defense and he was released.

Converse had other problems besides the election-day shooting. Stephen Gasters, county treasurer, unexplainedly dropped from sight in 1866. Converse was the last man known to have been in Gaster's company, and because of his previous record, he was suspected of being involved in the disappearance. Although no official action against Converse was ever taken, there were many whispered accusations by members of the community. The situation was compounded when Converse married Gaster's divorced wife in 1869, after being divorced himself by his first wife, Malvina. (Converse eventually proved that he hadn't murdered Gaster by locating him some three decades later as a prosperous old man in Nicaragua where he had run with the county treasury at the time of his disappearance.)

A 26-foot redwood log in the Converse Basin. Horse teams or sometimes even cavalry troops often posed on top of especially big logs. (Harold G. Schutt Collection)

New mill of The Sanger Lumber Company. Carriageway for logs adjoined the unloading platform and logs were pulled into the open framework structure in the center of the picture. (Dick Challacome Collection)

Feeling that he had been poorly treated, Converse withdrew with his new wife to the mountains south of the Kings River in the late 1860's and laid claim to a "large amphitheater of forest and chaparral encircled by mountain ridges" — the area now known as Converse Basin.

He announced an ambitious scheme to log the redwood trees by chuting them down to the Kings River and then floating them to a point in the valley below where a sawmill would be constructed.

Converse sent a few test logs down the river in an attempt to prove the feasibility of his plan and attracted the interest of several wealthy investors. The formation of the "Kings River Lumber Company" was announced in January, 1876, with Converse listed as a director, but a nationwide depression later the same year killed all hopes for the ambitious project.

Converse finally gave up the basin without obtaining title and took a job as a railroad lobbyist. For a time he was active in the mining business in Nevada and Utah. In later years, he presented himself as a "mining expert" in San Francisco where he was often on the verge of starvation. He became a lonely and bitter figure in the city.

His death came by drowning in San Francisco Bay and even this final incident was filled with question. Some said the drowning was accidental. Others claimed suicide, pointing out that Converse was a strong swimmer and that when his body was found, it was weighted with rocks.

In view of his many enemies, one would think that somebody might well have suggested a third possibility for his death — murder!

* * *

The difficult job of moving all milling equipment from the upper and lower mills to the basin named for Charles P. Converse was accomplished in the early spring of 1897. As there was no road from Millwood, everything was carried on railroad cars over Hoist Ridge.

The basin itself was a sort of elongated bowl, nearly 1000-feet deep, four miles long and ringed almost around with high ridges. The sloping sides were covered with great stands of fir, pine, cedar and redwood, broken only by a series of verdant meadows that stair-stepped to the basin floor. Here were thousands upon thousands of patriarchal forest giants stretching so high they almost blotted out the sky. Here was a sylvan paradise filled with colorful wildflowers, lush grasses and a variety of animal life. Some timber experts say that the Converse Basin contained the finest examples of Sierra redwoods to be found anywhere on earth. No less an authority than John Muir himself once examined and admired these magnificent trees. Aside from Charles Converse's brief stay and a few itinerant sheep herders, the basin had been virtually undisturbed for countless centuries. The tranquility was now at an end.

The new mill of The Sanger Lumber Company began cutting on June 30, 1897. The structure was known as a "two-band" mill, and all equipment was powered by a 1000-horsepower, single-cylinder steam engine. A crew of 400 men was employed as the season began.

The timber closest to the mill was the first to be cut, not only because it was handy but also to help clear the immediate area. Once on the ground, the trees were sawed into convenient lengths and dragged to the mill by donkey steam engines pulling steel cables. At the mill landing, a little cart mounted on a wide-gauge track lifted the logs up a runway to the cutting floor for splitting and sawing.

This procedure proved relatively easy so long as pine, cedar and the smaller redwoods were involved. But cutting down the giant redwoods — some of them 250-feet high and 25-feet thick —

Undercut in a Sierra redwood. Fallers stood on platform and often worked for a week or more to prepare the giant for the saw. (Harold G. Schutt Collection)

The long saws necessary to cut through the massive redwood trunks were made by brazing two regular saws together. Saws were sometimes 26 feet in length. (Harold G. Schutt Collection)

was an entirely different matter. Besides the difficulty involved in felling a tree that weighed several thousand tons, there was the additional problem of what to do with the massive trunk once it lay prone on the ground. In general, the method used to log the big redwoods was the same as for most smaller species, but the work was enormously more complicated, highly dangerous and very time consuming.

A platform was first built above the buttresses of the trunk, anywhere from five to 25 feet from ground level. Two men then climbed upon the scaffold each carrying a double-bitted axe. Using broad, swinging strokes, they gradually chopped their way to the very heart of the huge tree, sending great chips of wood flying 10 to 25 feet in the process. The rhythmic *whap — whap — whap* of the big axes resounded through the forest for days until a tremendous notch — six to eight feet high — had been cut from the trunk. The axmen took great pride in their ability with the double-bitted steel, and when they were finished with the undercut, it was almost as smooth as if it had been sawed.

Then, from the opposite side, a long cross-cut saw was drawn slowly back and forth across the trunk with an easy flowing movement. As the keen-toothed blade sank ever deeper toward the center of the tree, the two sawyers worked in such unison that they appeared to be a single fluid unit. A series of 24-inch steel wedges followed the saw to keep the weight of the tree from closing the path. The work was so arduous that even the hardiest crews required frequent relief.

After a week or more of labor, when only a narrow strip of the trunk remained between the saw and the undercut, the cutting stopped and more wedges were pounded into the gap, one after another. It took thousands of blows from the 12-pound sledges to finally force the tree from the perpendicular and start it toppling ever so slowly toward the chopped side, the uncut strip acting as a sort of hinge. Majestically, and with a mighty swaying and creaking, the big tree fell, gathering momentum as it hurtled toward the earth. The crash was like the roar of a raging sea beating upon a rocky shore. Dust and debris filled the air. The great stump, relieved of its burden of centuries, quivered for many minutes in a series of paroxysmal reactions. The ground trembled and shook as if struck by a mighty earthquake. A life, several thousands of years in the making, had been snuffed out in a moment.

Once the giant lay dead on the forest floor and quiet had again descended on the basin, the loggers rushed in, like Lilliputians toward Gulliver, to see the results of their handiwork.

At first, the destruction was beyond belief. Although the woodsmen were experts at their trade, felling the gargantuan redwoods was a new experience. Learning to do it successfully became largely a matter of trial and error. Many trees shattered into thousands of useless pieces when they hit the ground. The very size of the heavy trunks coupled with the extreme brittleness of the wood, made the redwoods highly susceptible to the slightest unevenness of slope.

"Lumbering of the big tree is destructive to a most unusual degree," commented an observer writing in the *Scientific American*. "The enormous size and weight entails considerable breakage when one falls. With The Sanger Lumber Company, it appears as if waste is the rule."

In time, a felling bed was devised that considerably lessened the damage. A crew of swamp-

The crash of a big tree. A shower of debris fills the forest as eager loggers look on. (Walter Smith Collection)

Giant bandsaw (redwood splitter) at Converse Mill. Johnny Holt, long-time employee, posed beside a big log. (Walter Smith Collection)

ers first cleared the area where it was planned for the tree to fall. The ground was leveled and all rocks, stumps and small trees removed. A cushion of branches, leaves and slash was then placed in position so that it hopefully would absorb much of the shock as several thousand tons of big tree hit the earth. Often it took nearly as long to prepare the bed as it did to fell the tree but the extra labor more than paid for itself in salvageable timber.

After the fallers had completed their work, the ground crews sliced the great trunks into logs which were then blasted apart using black powder packed into deep auger holes. This was necessary because the mill could not handle whole logs having the diameter of the larger redwoods. Sometimes the explosion split the timber more or less evenly, but often the brittle wood was blown into useless splinters. Eventually, a giant "redwood splitter" was added to the Converse mill equipment. This continuous, 90-foot ribbon of steel — called the longest bandsaw in the world at the time — could halve and quarter even the largest logs and eliminated some of the waste.

Inside the mill, the roar of machinery was deafening. The millmen relied on hand signals for the most part as the high-pitched screeching of the saws and the vibrant pounding of the steam engine made conversation impossible. As the rough-finish lumber emerged from the rear of the mill, it was loaded onto disconnected railroad log buggies and prepared for shipment up the incline, three sets of cars at a time. The end of the hoisting cable was attached to the lead car, and at the proper signal, the big winch at the top of Hoist Ridge began reeling in the steel line. The three-car train moved slowly up the 30 per cent grade. This was not a particularly steep pull for this type of hoist arrangement as inclines with grades up to 80 per cent have been used on similar operations. However, when compared to the normal railroad grade of one or two per cent, it was an abrupt ascension.

When the cars reached the summit, the cable was transferred to the opposite end of the trio of cars, and the load was then let down the Millwood side of the ridge to a junction with the regular rail line. When three loads had been sent over the tramway, the Shay locomotive tied onto the nine cars that were waiting at the foot of the hoist and pulled them over the narrow gauge rails to Millwood. Empty cars were shipped back to the mill using the same procedure in reverse.

The hoist was powered by two large steam boilers. It had a tremendous drum, 26 feet in diameter, which held more than two miles of $1\frac{1}{2}$ inch cable in a single layer. Since the cable did not pile up, an unvaried speed could be maintained on either side of the ridge. As with most

Wreck near the top of the Converse Hoist. Mill can be seen at right of picture. There were many such accidents on the incline. (Paul Spivey Collection)

ABOVE: Hoist house atop Hoist Ridge facing the Millwood side. Here cable was transferred from one end of load to the other. (Walter Smith Collection)

BELOW: Interior of hoist house. Giant drum which held the long cable in a single layer can be seen at upper right. (Walter Smith Collection)

ABOVE: Good shot of tramway and front of mill. Redwood splitter was to left of log carriage in the section of mill that juts out. (Walter Smith Collection)

BELOW: Rear of mill where cars were loaded for the incline trip. Arc lights provided some night lighting. (Bill Reid Collection)

early inclines, signals were given to the hoist engineer by means of a telegraph line paralleling the route. By shorting the bare wires with a metal-tipped pole, directions for stopping or starting were transmitted using prearranged signals.

Despite the most careful precautions, there were many wrecks and derailments in the hoisting operation. The story is told by old timers that the hoist engineer was offered a new suit of clothes by the company if he could run the incline two full weeks without an accident of any kind. No one ever collected the bonus!

At Millwood, great stacks of lumber filled the drying yards. Busy flume crews performed their clamping and shipping tasks from dawn to dusk. On Saturday nights, Millwood's pleasure palaces boasted capacity business as thirsty loggers flocked into town from the basin on special "passenger trains." Tourists from as far away as San Francisco were seen hiking, fishing and riding the trails in the invigorating Sierra air.

"Bloomers are out in force this year," reported the Visalia *Delta* on June 10, 1897. "The tourist season has started early and the hotels are very busy. There were 70 guests at the White House alone last week."

Millwood was little affected by the moving of the mills to Converse. Its function as a resort, supply center and shipping outlet seemed to assure its prosperity for the immediate future.

* * *

Maintenance of the flume was a considerable concern for the Sanger Lumber Company. The big trough required constant upkeep as leaks or jam-ups could result in costly shutdowns. On the lower end of the route, provisions for the lonely flume herders were hauled by wagon from Sanger. On the upper portion, supplies were sent by flume boat from Millwood.

Flume boats were constructed in an assortment of styles and sizes depending on the whim of the builder. Generally they were V-shaped to conform with the trough itself, and usually from 12 to 20 feet in length. The rear end of the "V" was closed, but the front end was left open so water could enter and hold down the boat on the curves. A flat board across the top held both passengers and supplies.

No one was ever seriously hurt riding the Millwood flume, although two men were dumped out of their boat on one occasion and had their luggage thoroughly soaked by water. The company doctor rode the flume one night on an emergency case. Another time, a nurse brought an expectant mother down the flume to the Kings River where they were transferred to a wagon for the remainder of the trip to Sanger. Lumberjacks often used flume boats on Sundays to ride to the river for a day of trout fishing. On such excursions, the eager anglers faced a 10-mile, uphill hike back to town. Flume boats were returned from Sanger to Millwood by wagon with an upcoming load of clamps or supplies.

Flume boats like these were common on the route. Boys appear to have outrun the water. (Don DeVere Collection)

A ride on the flume was called "one of the greatest pleasures in the world" by the author of an article appearing in an 1897 issue of the Visalia *Times.* The writer, Edmund Norton, made the trip along with five other passengers and gives a vivid account of his experience.

The start was made at 3 p.m. after the last lumber had been placed in the flume for the day, carrying with it the familiar "joker" that signified the final shipment. A 16-foot boat was then placed in the flume and nailed to the side while

Two examples of feminine flume activity. Woman in flume boat (above) seems quite sedate for such a circumstance. Girls (below) have a rather long, narrow swimming pool. (Dick Challacome Collection)

it was being loaded. Into the boat were placed two 100-pound kegs of nails, a box of groceries, a bag of barley and several other articles for the various stations along the way. The six men then climbed aboard and the nails were pulled releasing the boat.

"There is a splash and a gurgle as the suspended end drops into the water. A slip and the next moment you are going a mile every three or four minutes. As we approached the lower mill, the boat hit a steep pitch in which gravity took over and caused it to drop faster than the water. This slide was one of the great thrills of the trip."

The boat stopped at the mill and three men, whose destination it was, got off. Below the mill was Devil's Slide, a place where the flume dropped sharply for two miles and made a series of sharp curves. Next were three trestles, all more than 100 feet in height, which presented such wonderful opportunities for scenic views that "no thought was given to the possible dangers." Lower down, at flume station Number Four, the boat was exchanged for a flat raft with box seats for the trip into Sanger.

Norton was most impressed by the never-ending maze of changing mountain scenery. "In eight or ten hours you have passed through all the gradations of the season!"

In more ways than one, a ride in a flume boat was a never-to-be-forgotten experience!

* * *

The first season at Converse Basin was a complete disaster for The Sanger Lumber Company. Any hopes that Hiram Smith entertained for reversing his previous misfortune vanished after only two months.

"Hardly a day goes by at Converse without a breakdown," reported the Visalia *Delta* in July, 1897. "There is a great lack of steam power although nine furnaces are kept going all the time. Fuel is largely wet sawdust and green wood. The loggers are having great difficulty handling the big trees and there are many accidents. Two men were killed this week. Cutting is much slower and considerably less than was done previously at the two mills."

A lumber load with passengers is lowered down the Millwood side of the hoist. From here the Shay locomotive will carry the car to the drying yards at Millwood. (Mac Grosse Collection)

Down the Converse Hoist

Accidents had plagued the Sanger operation even before the move to Converse. Logging in the Sierra was a dangerous occupation and the slightest miscalculation could bring sudden misfortune in a hundred different ways. Exactly how many men were injured or killed over the years is anybody's guess, but some of the typical accidents were described in contemporary newspaper stories:

In 1891, William Stevens, a logger, was badly hurt when a carload of logs bound for the lower mill jumped the railroad tracks and rolled down an embankment on which Stevens was standing. He eventually recovered.

In 1892, a fire at the lower mill burned a dozen cabins. No one was seriously injured but one lumberjack lost $40 in gold which melted in his trunk.

A carload of logs jumped off the Converse hoist in 1893 and badly damaged the track. One car was smashed to pieces but two men riding on the car escaped injury by jumping in the nick of time.

Another hoist accident took the life of a Dane identified only as "Werskind" in 1895. Four cars of timber were being lowered down the incline with eight men aboard the load. The drawhead connecting the cable to the donkey engine came loose and the cars flew free. Everyone except the unlucky Werskind jumped. He was crushed in the subsequent crash.

On July 18, 1895, a huge flywheel which powered the bandsaw at the upper mill flew into pieces. The roof of the mill was peppered with holes and a "hailstorm of debris" fell on the workers. No serious injuries were reported but damage of $3000 was caused to the equipment.

In 1896, the deputy county clerk, seeking a free train ride, attempted to jump onto the Shay locomotive as it moved slowly down the tracks near Millwood. He missed and had two toes crushed for his trouble.

Breaks in the flume were common. Here a crew repairs a damaged section. (Jean Hume Browning Collection)

On July 23, 1896, Charles Pierson, a Swedish logger, was crushed to death by a log when a skid hook came loose. The report of the incident contains the macabre fact that Pierson's coffin was constructed of lumber from the very log that killed him!

There were other problems besides those which resulted directly from the logging activity. Several Millwood residents contracted smallpox in 1894 and it was feared for a time that an epidemic might sweep the community. Fortunately, the disease was confined to the original victims who soon recovered their health.

Frank Davenport, alleged train robber and associate of Chris Evans, was arrested in a Millwood saloon in 1896. It was rumored that the sheriff had been tipped off by a local informant as to Davenport's whereabouts and there was some worry among the more timid residents that Davenport might return to Millwood in a belligerent humor. The concern proved to be ill-founded.

On August 30, 1896, one Cass Colvin murdered his brother-in-law, R. L. Totten, by slashing his throat with a razor after an argument about rent owed Totten's wife who ran a Millwood boarding house. Colvin, who had a reputation as a troublemaker, hid among the piles of lumber for three days while county officers and United States cavalrymen searched the woods. He finally turned himself in to the officers and was spirited off to jail in Fresno before the angry loggers were aware of his presence. It was the sheriff's candid opinion that they would have hung him on the spot. Colvin was later sentenced to life imprisonment at Folsom penitentiary.

* * *

The total cut for the year of 1897 was a poor 16,000,000 board feet of lumber — hardly enough to justify the expense of the move to Converse. Faced with ouster as manager, and beset with lawsuits and debts on every hand, Hiram Smith finally abandoned his sinking financial ship and filed for bankruptcy with the Fresno courts on August 21, 1897. He listed liabilities of $798,138.13, with no assets.

"The largest insolvency suit in the history of the county was that filed by H. C. Smith, who was one of the former firm of Moore and Smith," commented the Fresno Morning *Republican.* "The true amount of Smith's failure is not so large as that indicated by his statement as the heavier claims were contracted jointly by him and his former partner, A. D. Moore. However the petitioner is liable for the whole amount in the inability of Moore to pay, but the fact is that Smith is responsible for only half of the indebtedness. Smith desires to be freed from every liability and hence the large amount stated."

The majority of the 34 persons and firms making up the long list of creditors had taken The Sanger Lumber Company stock as collateral for their loans to Smith and Moore during the 1894 reorganization. This group included several banks and business houses as well as many individuals. The principal creditor was the San Francisco branch of the Canadian Bank of Commerce, and it was the management of this bank which had assumed the position of assignee for the company in 1895 when Smith and Moore failed to meet their obligations.

Another view of lumber cars coming down the hoist. Usually three cars were handled at a time.

The bank manager, George H. Collins, and later, Archibald A. Kains, served as titular head of The Sanger Lumber Company with the title of president. Frank Boole, an experienced California lumberman, was hired to replace Smith as the actual manager with headquarters at Sanger. Although the bank was certainly not anxious to continue in the lumber business, the only opportunity to realize any reasonable return on the tremendous investment in the mill and flume was to attempt to sell the operation as a going concern. Even as Boole took charge, bank officials made plans to unload their unwelcome white elephant at the earliest opportunity. This proved to be a more difficult task than was at first anticipated.

The unfortunate Hiram Smith left the Fresno County area but not the lumber business. With friends and relatives, he became involved in sev-

ABOVE: View of locomotive with lumber load at Millwood yards. Shay has backed down a trestle to reach the proper area. (Jean Hume Browning Collection). BELOW: Bucking crews ready a huge log for the chute. Handling the gigantic loads took great skill and courage. (Harold G. Schutt Collection)

eral logging companies in northern California including the Pacific Lumber Company and the Excelsior Redwood Company. He later was associated with logging operations in Mexico, Brazil and British Columbia.

* * *

Frank Boole gradually instituted many changes in the Converse logging show. He constructed an ingenious series of great chuteways extending several miles from the mill so that outlying stands of timber could be logged.

The chutes were essentially a continuous line of two parallel logs, each about 18 inches thick, and placed from one to two feet apart. A third timber was sometimes sunk in the ground between the two slide timbers for additional support. The parallel logs were cut from trees having a minimum taper, usually 20 to 60 feet long, and were laid with their butts up-grade which made them less apt to sliver. A simple lap joint connected the chute sections. On level stretches, the slide was built directly on the ground and required a minimum of bracing or support. On steep pitches and in crossing depressions, the chute was supported by built-up crib work or securely anchored trestles as rigidity was all important. Logs as thick as 10 feet in diameter were sometimes used for foundations in the trestle work. The tops of the chute timbers were hewn smooth with a broadax and slanted slightly inward to form a cradle over which logs could be dragged. Whenever the chutes curved, oaken rollers or upright "roller logs" were positioned in the arc of the curve to keep the pulling cables on line. In the same manner, road rollers were fitted between the chute timbers that spanned the brows of hills.

Two main-line chutes terminated at the basin mill. An additional network of feeder lines switched into these main chutes so that logs could be conveyed from almost any point in the basin area. Donkey yarding engines stationed at strategic points along the route powered the feeder chutes. Two huge permanent hoists, one on either side of the mill, provided the pulling force for the big chutes.

At the cutting areas, bucking crews set out the saw timber — logs 10 to 24 feet long. These were maneuvered down the feeder chutes onto the main line and "dogged" together by means of hooks and huge chains. Blasting was required only on logs that were too uneven to move smoothly over the slide. Log after log was lined up by the rigging crews until a continuous train — sometimes more than 300 feet in length — had been formed. The main cable, a wire rope $1\frac{1}{4}$ inches in diameter, was then fastened to the trailing log. Signals were given to the hoist operator in the basin, a mile or more distant, and the big bull donkey at the mill began to slowly pull the ponderous train in from the woods.

To facilitate the progress of the gigantic load, "greasemonkeys" kept the chute timbers well swabbed with animal fat or machine oil to reduce the heat generated by the heavy logs as they were dragged along the wooden slide. Showers of sparks erupted from the train as it made its majestic way toward the mill. The big logs themselves often smoked from the friction induced by the rubbing of timber against timber.

Perched daringly atop the last pusher log in the train was the "trail rider" or "log conductor" who had the difficult and dangerous job of instructing the unseen hoist engineer in his operation of the cable. By means of a coded bell system, the trail rider regulated the movement of

Excellent picture of a log train being prepared for a trip down the chute. (Jean Hume Browning Collection)

the train by shorting with a metal-tipped pole the bare telegraph wires that paralleled the route. The hoist man, working with both hands on the control levers, had only these signals and the feel of the pulling cable to guide him. The hour-long trip was tedious and fraught with peril every step of the way. Sometimes a log would suddenly rise midway up the line and ride out of the trough, its tremendous weight carrying it crashing down the slope like an enraged bull as it rolled over everything in its path. On other occasions, the entire train would buckle. Logs flew up from the slide like ten pins in a bowling alley, often setting themselves and the chute on fire as they parted from the course and splintered into kindling wood against the canyon walls.

Two steel cables were used on the chuteways. The heavy main line pulled the train itself, while the smaller "back line," hanging from cables on trees wherever necessary, ran directly over the hills from the main hoist to the origin point thus forming a continuous loop. The back line was used to return the main cable to the cutting area along with the dogs, chains and other chuting apparatus. On the return trip, the trail rider was pulled in a "chute boat" — a hollowed canoe-type log. The haul-back trip was even more hazardous than the downward journey despite the absence of the massive log load. At each curve, the heavy main line would slap across the chute boat like an angry whip as it straightened itself out, and the rider had to jump at precisely the right moment

Panoramic shot of the Converse Mill operation. This gives an excellent idea of the function of the log chutes and the incline tramway. At left and right of picture are hoist houses that held the giant, main-line chute hoists.

Below is a side-angle view looking over the incline into the mill. Big log splitter is housed in the front portion of the mill. Tramway runs through the mill at the right. (Harold G. Schutt Collection)

or be cut to pieces. He probably looked much like a modern surfboarder as he balanced precariously on his hollowed-out perch and rode his boat over the hills and around the curves back to the uncut forest area.

The chute rider was unquestionably the greatest daredevil of all in a profession where every task was filled with danger. The lusty, brawny crew of 700 men who made up the work force at Converse Basin was a breed apart from the ordinary. Many nationalities were represented — big, raw-boned Swedes; husky Irishmen; sturdy Poles, Danes, Norwegians and Germans — strong men who put great store in their own ability with axe, saw or fists. Here were men who could labor arduously for 11 hours a day, six days a week, and still hike 20 miles on Sunday in search of a new fishing hole. The Swedes seemed to have a special ability in the lumberjack's art. A popular saying of the day when a particularly tough job needed doing, was "send me six men or two Swedes."

The Converse loggers earned every cent of their pay — $1.50 to $2.50 per day, less board for single men. There were no unions, no compensation laws and few safety regulations. An injury was an individual matter between the employee and the management, and seldom could the unlucky woodsman afford a costly lawsuit to recover damages. Accidents occurred in every phase of the operation and the rustic hospital at the basin was well filled during most of the season.

Two excellent pictures of logs moving to the mill by means of chutes. Mill is visible in top photo as a short log train is pulled toward the landing. A trio of trail riders stands atop log in the photo below as a cloud of dust fills the air. (Both Harold G. Schutt Collection)

Fallers faced the threat of "backlash" whenever they downed a big tree. Sometimes the mighty redwood strayed slightly from its intended downward course and crashed against smaller trees in its path, bending them over like limber fishing rods. As the small trees snapped back to their normal positions, the unfortunate crew was often showered with a barrage of heavy, projectile-like limbs.

The mill workers had their own special worries. All day long, the carriage men fed a continuous stream of logs into the two small bandsaws and the big redwood splitter. If the slender bands of steel struck any foreign object such as an imbedded rock or hook, the result was like the explosion of a shrapnel grenade. Chunks of steel and wood rifled through the confining quarters of the building with the force of a hail of bullets.

The greatest danger of all was from the miles of steel cable that threaded the basin. The chute men, swampers and fellers were contantly warning each other to stay clear of the "bight of the line." A "bight" is a curve or bend in a slack line as it lies without tension on the ground. When the steam donkey operator, often some distance away and out of sight, began applying pressure to his line, the cable was pulled taut with a sudden jerk, and woe to the unlucky woodsman who stood in its path. At full tension, the steel line could cut a man in two like the snapping of a twig.

But the hazards involved in a day's work were all part of the game to the carefree lumberjacks. Their energy was boundless and their desire for pleasure insatiable. It was no surprise that a saloon and gambling hall soon occupied a prominent place in the little community that adjoined the Converse mill. The men no longer made the train ride to Millwood in search of entertainment.

On weekends, gambling games lasted nonstop from Saturday evening until the whistle sounded the call to work on the following Monday morning. Family groups sought more subdued amusements of their own in the form of square dances, polkas and jigs at the community dance hall.

* * *

Each winter, when mountain snows closed the Converse mill, the lumberjacks scattered their separate ways. Some had off-season jobs; others passed the time in idleness. In either case, the loggers somehow found their way back to the basin with the coming of spring, eager to sign on for another season. There was the usual turnover of personnel each year — a common occurrence in lumber camps — but new hands quickly replaced those who moved on to what they hoped would be greener pastures.

During the off-season, the Converse woodsmen took great delight in spinning yarns about the size of the big trees they were logging at the basin. A favorite story told about a group of five or six men who were felling an exceptionally big tree. After a week of work, they began to hear faint tapping noises. The following day was a Sunday and they decided to explore the source of the

Big bull hoist in shed near Converse Mill. This engine powered the main line chute east of the mill itself. (Harold G. Schutt Collection)

A combination of donkey engine and horse teams is used to handle logs in this Converse chuting operation. (Harold G. Schutt Collection)

ABOVE: The small Abbott Mill (Camp Three) which was built in 1902 to handle timber on the Millwood side of Hoist Ridge. (Dick Challacombe Collection)

■

strange noises. They arose very early and set out on an expedition around the tree they were felling and by noon, had reached the opposite side. Here they found another crew of four or five men felling the same tree — the source of the faint tapping noises!

Apart from the expedition around the tree, the story is hardly exaggerated.

* * *

The Converse logging show of The Sanger Lumber Company lasted for slightly less than a decade. During that period, cutting was conducted at various times out of five separate lumber camps. These outlying forest "villages" — consisting of crude buildings with no modern conveniences — were used to headquarter the crews as they moved from one remote area of uncut timber to another.

There were three camps in the basin proper (Camps One, Four and Five). From these operations, logs were fed to the mill over main-line chutes. Two other bases (Camps Two and Three) were located in territory adjacent to the basin itself. Camp Three, also called the Upper Abbott Mill, was built in 1902 to cut timber that had originally been bypassed in the move to Converse Basin. It was situated about a mile below the foot of the Converse Hoist, on the Millwood side, immediately contiguous to the railroad. Camp Three was a very small mill and processed mostly pine of a moderate diameter. The building of this mill made it possible to handle timber on the Millwood side of Hoist Ridge without hauling it back and forth over the incline.

By far the most interesting, and certainly the most challenging of all the many chuting operations was that which originated in 1901 at Camp Two. This logging chance was established in a well-timbered valley known as Indian Basin, directly east of the Converse Mill on the other side of a high ridge called Rob Roy Mountain. Bringing logs to the mill over the 500-foot rock barrier posed a difficult problem. After considerable study, a breathtaking chute was constructed running directly up the side of the ridge from Camp Two to the summit. Grades of nearly 70 per cent were used in places. A large steam hoist, set on bedrock at the top, powered the cable which pulled the heavy log trains up the hillside chute. At the crest, the trains were transferred to a steel line connecting with one of the permanent bull donkeys at the basin mill. The unique arrangement worked surprisingly well considering the difficulties involved, and a goodly amount of timber was carried over the Rob Roy Hoist during its several years of operation.

Installation of the big bull hoist engine at the bottom of main-line chute to Camp One. This engine can be seen in shed at left of photo on Page 72 (top). (Grover Johnson Collection)

* * *

About 1903, when the loggers were beginning to probe the extreme north end of Converse Basin, they came upon an ancient redwood that dwarfed all others they had ever seen. Word of the discovery flashed quickly through the camps! On the next few Sundays, nearly every lumberjack in the basin journeyed out to see the massive, scarred old relic for himself.

"Church" established in Converse Basin in 1903. Man second from left in top row is Jack Clark, later mill superintendent, who was of great help in the preparation of this book. (Walter Smith Collection)

Many thought it was surely the very largest tree in the world; all were certain that it was by far the biggest in the basin. A. R. Moore, who had a photo studio at Millwood, spent the next few weekends taking pictures of loggers posing beside the tree.

The management at Sanger soon heard about the huge redwood which stood directly in the logging path. "Can we get it out?" was the first question asked by company officials — a typical lumberman's callous reaction to news of such a discovery.

What happened after that is not exactly clear. Only one thing is really certain: the monstrous tree was never cut. Some stories say that it was simply too big to be profitably logged. With plenty of smaller, easier-to-fell trees all around, there was no need to bother with this one. Others maintain that the United States Bureau of Forestry, prompted by conservationists, arranged a transfer of nearby timber in exchange for saving the redwood giant. Whatever the case, the fact remains that the tree was spared — the only really big tree left uncut in the entire Converse Basin.

Somebody, and again no one is certain who, gave it the name "Boole" after Frank Boole, general manager of The Sanger Lumber Company. Whether Boole actually had any particular role in preserving the grand old redwood, or whether the attaching of his name to the tree was merely a political move by conservation enthusiasts, is at this point academic. The name stuck.

For many years, the Boole was considered to be the largest tree in the world. In 1928, accurate measurements of all the big redwoods were taken, and the Boole, although largest in circumference, was found to be third in total bulk, only slightly behind the Sherman and Grant Trees of nearby Sequoia National Park.

The Boole Tree measures 269 feet in height, 112 feet in circumference and 35 feet in base diameter — truly a magnificent redwood made even more impressive by the solitary splendor of its location. It still stands today in Converse Basin, a lonely relic of a long-ago age.

* * *

While serving as assignee for The Sanger Lumber Company, officials of the Canadian Bank of Commerce persisted in attempts to dispose of their unwanted operation. At different periods, several prospective buyers evinced a mild interest in the property including Elmer Cox, superintendent of the Madera Sugar Pine Company, but no

The Boole Tree. (Jean Hume Browning Collection)

Summer in the Sierra---a Long Time Ago

Loggers and their families posed at the bottom of the Millwood side of the Converse Hoist. Shay locomotive picked up lumber cars here for the trip to the drying yards over the narrow gauge railroad. (Stanley Snook Collection)

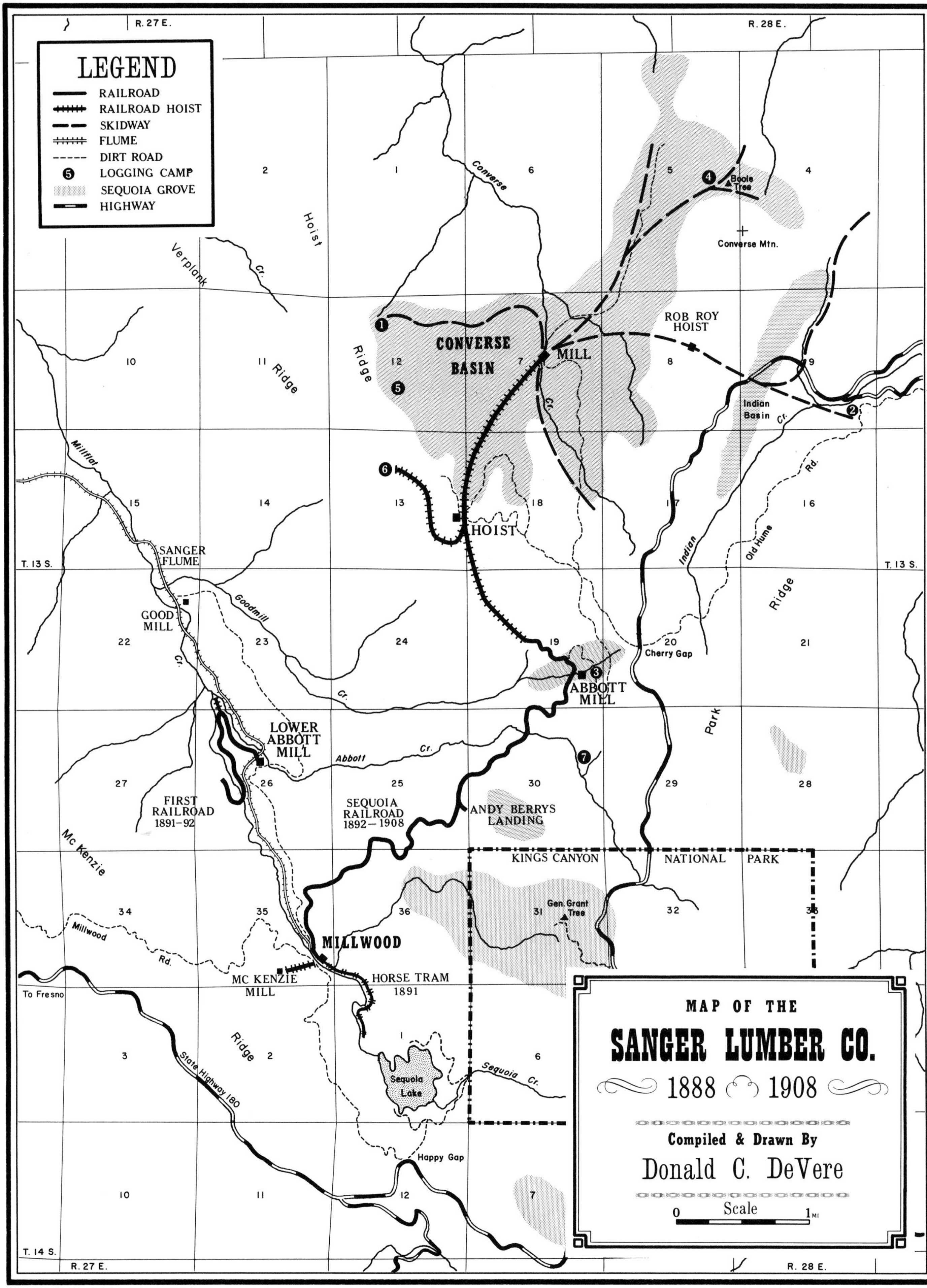
LEGEND
RAILROAD
RAILROAD HOIST
SKIDWAY
FLUME
DIRT ROAD
LOGGING CAMP
SEQUOIA GROVE
HIGHWAY
MAP OF THE
SANGER LUMBER CO.
1888 1908
Compiled & Drawn By
Donald C. DeVere
Scale
0
1 MI
R. 27 E.
R. 28 E.
T. 13 S.
T. 14 S.
CONVERSE BASIN
MILL
ROB ROY HOIST
Converse Mtn.
Boole Tree
Indian Basin
HOIST
SANGER FLUME
GOOD MILL
ABBOTT MILL
LOWER ABBOTT MILL
FIRST RAILROAD 1891-92
SEQUOIA RAILROAD 1892—1908
ANDY BERRYS LANDING
Cherry Gap
KINGS CANYON NATIONAL PARK
Gen. Grant Tree
MILLWOOD
MC KENZIE MILL
HORSE TRAM 1891
Sequoia Lake
Happy Gap
To Fresno
State Highway 180
Verplank
Hoist
Converse
Millflat
Goodmill
Abbott Cr.
Mc Kenzie
Millwood Rd.
Ridge
Sequoia Cr.
Indian Basin Cr.
Old Hume Rd.
Park Ridge

reasonable offers resulted from the overtures.

Finally, in 1905, a sale appeared to be in the making. The interested party was one Ira Bennett, general manager of the American Lumber Company of Albuquerque, New Mexico, and an experienced lumber dealer. On June 14, 1905, in return for a $1000 deposit, Bennett was given a six-month's option to purchase all physical assets of The Sanger Lumber Company including the yards, mills, lands and flume. Bennett was not a wealthy man by any means, but he was a capable promoter. He immediately set about the task of arranging the financial backing needed to consummate the deal in the required time.

He found that backing in the person of Thomas Hume, wealthy Muskegon, Michigan lumberman and industrialist, who owned timber properties in nine states and Canada. Hume was anxious to replace his cut-over Michigan forest lands with more fertile western property, and sent investigators to look into Bennett's proposal. Their reports concerning the timber were favorable.

While negotiations were reaching fruition, a disastrous fire broke out in the Converse mill at 1 a.m. on November 11, 1905, and burned the structure to the ground. There was little loss of lumber, but damage to the mill was estimated at $100,000. Since the physical plant represented by the mill building was only a minor part of the whole package, the fire did not affect the sale of the company.

On December 12, 1905, Bennett's option was exercised and the assets of The Sanger Lumber Company were sold to a combine headed by Thomas Hume. It was said that the new owners planned to move the logging operations to virgin timber farther to the east. Instead, the Hume forces continued at the basin for almost two full seasons (see next chapter) but it was mostly a "mopping up" activity. To all intents and purposes, the Converse logging show was over.

It had been by far the single most destructive force to attack the *Sequoia gigantea* in the memory of modern man — worse by far than all the fires, storms, winds or floods that nature could muster. Exactly how many of the mighty redwoods were destroyed in and near the basin will probably never be known. Some experts say the total might have exceeded 8000 trees!

The most ironical part of the whole affair is that it was all for naught. Not one cent of profit was apparently ever realized by those responsible for the destruction. The instigators, Smith and

The rustic hospital at the basin. Injured loggers are much in evidence. (Walter Smith Collection)

Moore, found only bankruptcy. The abettors, the bank and other creditors, suffered untold losses. Exactly how much money was poured into the ill-gotten enterprise can only be surmised, but Frank Sayre, long-time bookkeeper at the Sanger plant, states unequivocally that each season's operation resulted in a deficit of "many thousands of dollars."

During the nine years of cutting at the basin, The Sanger Lumber Company milled a total of about 191,000,000 board feet of lumber, an average of 21,200,000 board feet per year. The greatest cut was in 1903 when 26,177,936 board feet were processed. The worst season was 1897 when only 16,000,000 feet went through the mill. From one-third to one-half of the total yearly cut was redwood, depending on the area being logged at the time.

Why was there no profit? Probably because of the enormous complication of the operation. To fell big trees in rough terrain, chute them to the mill for cutting, and then haul rough lumber over the hoist to Millwood for fluming represented a tremendous expense. The flume itself was a ponderous method of transporting lumber when compared to that used by other less remote milling operations.

Waste was also an important contributor to the overhead. Waste of time in felling timber that shattered beyond salvage; waste in blasting

Timber was felled all around the Boole Tree. Here a donkey crew pauses for the photographer in front of the giant redwood. (Jean Hume Browning Collection)

logs that split into kindling. Waste in nearly every facet of the operation.

As Gifford Pinchot, Chief United States Forester, observed in 1903: "It is true that lumbermen in America are second to none in skill and ingenuity, in the perfection of their tools, and in the effectiveness of the methods they have devised. However, it is unfortunately also true that the present methods of lumbering are tremendously destructive, wasteful and costly to the operator in the end." Pinchot listed The Sanger Lumber Company as a prime example of one that needed much improvement in the field of conservation.

Perhaps the most telling indictment of the logging done at Converse Basin was that made in 1930 by John R. White, former Superintendent of Sequoia National Park, and Walter Fry, United States Commissioner for Sequoia National Park, in their book *Big Trees*. They wrote of an early day visit: "We entered the Converse Basin with all the eagerness of antiquarians about to visit the ruins of a vanished race . . . Hour after hour we tramped beside fallen columns more massive than those of Grecian temples; or climbed on stump after stump that measured over 20 feet in diameter at several feet above the ground, stumps that had once been part of the great trees, 30 feet in diameter and nearly 300 feet high. And always we marveled at the relatively little timber of the big trees that had been worked into lumber . . . Scores and scores of trunks remained on the ground, of which not one-fifth of what was commercially valuable had been used. That is a conservative estimate. There were fallen trees which had hardly been touched after they had been felled; perhaps because they lay awkwardly or were too big to handle . . . There must be millions and millions of feet of lumber in the boles and limbs of sequoias lying on the ground in the Converse Basin; and there are millions more in the chutes and trestles on which the logs were conveyed to the mills. There are miles of such trestles and chutes — miles and miles and miles. Waste, waste, waste. Probably not one-half of the trees destroyed, of all species, ever reached the mill to be converted into lumber; quite possibly not one-third. The

Below the crossing of the Kings River near Trimmer Springs, the flume paralleled the river on its way to Sanger. (Bill Reid Collection)

Two donkeys being brought down the incline tramway. Big bull engine shown on left and smaller upright type at right. Man in suit and derby is Dr. Sweeney of Fresno who served as company doctor. (Harold G. Schutt Collection)

destruction of that mighty forest did nobody much good, but has deprived the people of America for all time to come, of a national playground equal or superior to existing National Parks."

This is strong language. However, in fairness to the woodsmen, attitudes had changed in the intervening years between the Converse lumbering and the visit of Fry and White. The early day loggers lived in an America that seemed to have resources without end. The cry was loud for lumber to build the towns and cities of a growing west. The feeling for conservation was yet to come. When forests were cut over, the loggers simply moved over the hill to a new timber area. And then over the next hill, and the next. The redwoods were merely trees to be cut like any other, except that they were bigger. Who thought there could be any end?

To their everlasting credit, some men were wise enough to see what the future held. Through their influence, the *Sequoia gigantea* were saved in Yosemite, Sequoia and other groves that the public now enjoys and admires.

Teddy Roosevelt once said, "There is nothing more practical in the end than the preservation of beauty."

In the matter of the Converse Basin, it is unfortunately too late.

A favorite picture spot was this trestle on the Sequoia Railroad, midway between Millwood and the basin. Here are two different, but similar shots of lumber trains. (Dick Challacombe Collection)

Sequoia Railroad

The narrow gauge rail line that wound its way through the timber hauling logs and lumber was no great main-line railroad by any means. It was built of very light, 30-pound iron and had only one engine — a 28-ton, two-truck Shay. Pictured at left is a train load of lumber going down grade on its way across one of the 23 trestles between Converse and Millwood. (Dick Challacombe Collection). Below is a marvelous early day view of the wood burning Number One at the foot of Hoist Ridge. (Jean Hume Browning Collection). Opposite page shows the railroad terminus at Millwood. The railroad entered this area from the left of the picture. Peaked roof building in center was a combination store and warehouse. Excellent map by Don DeVere depicts the railroad unloading spur and the handcar trestle layout. Note how flume converges from various starting points for trip to Sanger. (Photo Stanley Snook Collection)

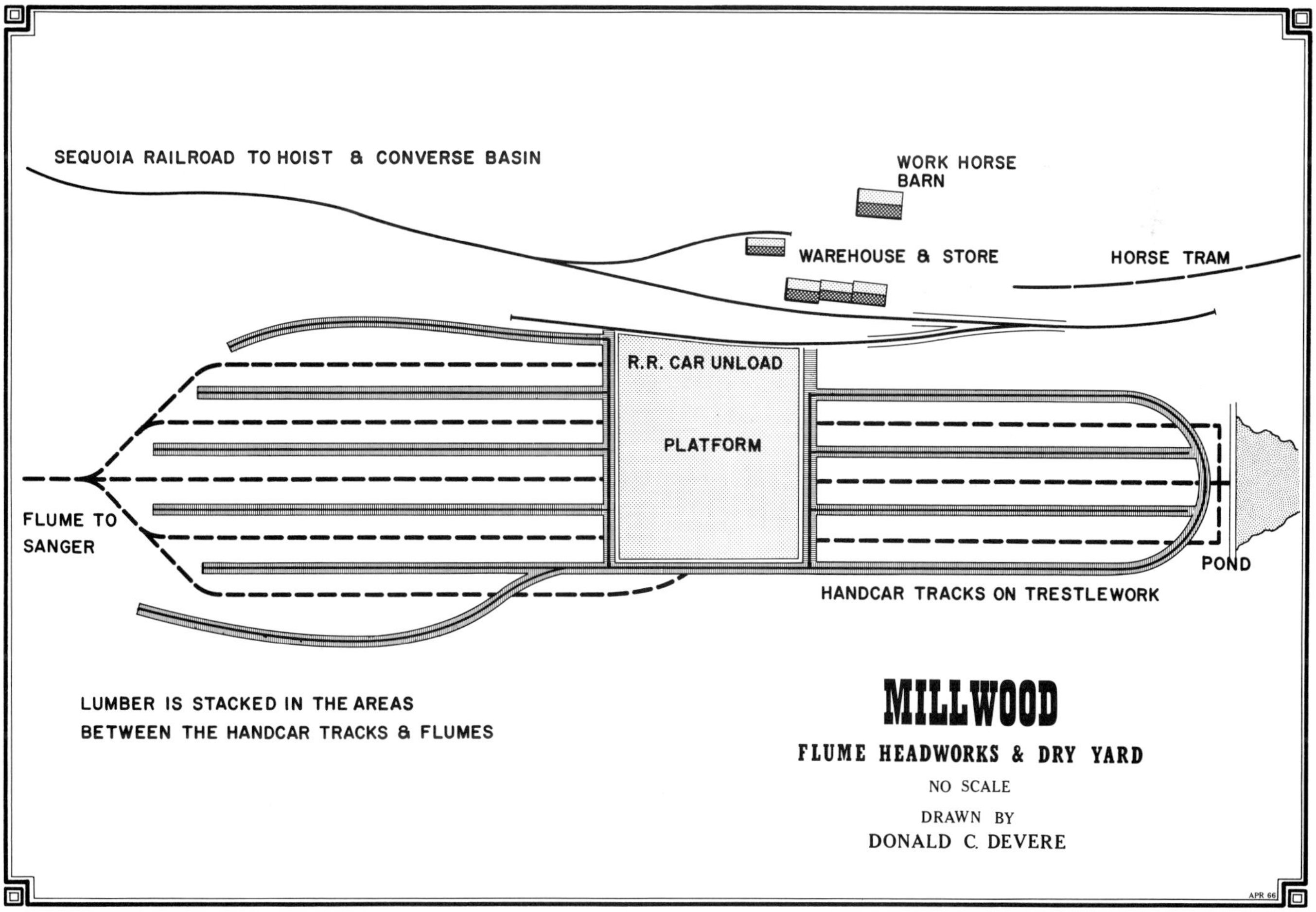
SEQUOIA RAILROAD TO HOIST & CONVERSE BASIN
WORK HORSE BARN
WAREHOUSE & STORE
HORSE TRAM
R.R. CAR UNLOAD
PLATFORM
FLUME TO SANGER
POND
HANDCAR TRACKS ON TRESTLEWORK
LUMBER IS STACKED IN THE AREAS BETWEEN THE HANDCAR TRACKS & FLUMES
MILLWOOD
FLUME HEADWORKS & DRY YARD
NO SCALE
DRAWN BY
DONALD C. DEVERE
APR 66

BELOW: Bucking crew readies a giant redwood log for the chute. A total of 25,000 board feet is marked on the log which would make it contain enough lumber to built a house! (Harold G. Schutt Collection)

ABOVE: Explosion at Camp Four on June 2, 1905. A cache of black powder was stored in a building that caught on fire. Fortunately, the crew was given sufficient warning and was able to clear the immediate area before the powder explod-

ed. No one was injured but the camp was badly damaged as shown in this view taken shortly after the episode. Danger was ever present in the logging camps. (Harold G. Schutt Collection)

BELOW: The Admiral Dewey Tree in Converse Basin was variously listed as being from 18 to 26 feet in diameter. Here is a large section of the trunk after the unfortunate giant was felled for lumber. (Harold G. Schutt Collection)

CHAMPION FALLERS

Although almost all of the axmen employed at the basin were proficient with the heavy steel, several were acknowledged to be champions at handling the tools of the trade. Big Bill Mills (shown above right) stood well over six feet, weighed 230 pounds or more, and could do the work of two men. He was strong as an ox and had a hand in felling more of the big trees than any other single person. His son-in-law, Bill Fauts, is the other man in the picture. At left is Mills again with Fred Jarrett, another champion faller. It took a special breed of man to swing the mighty broadax for hours on end. The fallers were literally the kings of the woods and tales of their prowess were favorite topics whenever loggers gathered to spin yarns. (Harold G. Schutt Collection)

PICTURE TREES

A favorite pastime of the fallers was to cut an especially large undercut in a huge tree and then pose for pictures when the photographer came around on the following Sunday. Here are two of many such existing pictures.

CHUTES

As mentioned in the text, a network of chuteways combed the basin bringing in logs from the outlying camps. Buckers (above) are preparing to blast an uneven log before placing it in position on the chute. Another crew with big log poses below. (Both Harold G. Schutt Collection)

ABOVE: Long train of logs being readied for a trip down the chute to the mill. Steam yarding engines drag the logs into position.

BELOW: Part way down a chute. Log has evidently been blasted and yarder is hauling it to the main line. (Both Harold G. Schutt Collection)

ABOVE: Filing and sharpening the big band saws was an art in itself. Here the crew of the filing room pauses for a group picture. (Grover Johnson Collection)

BELOW: Loggers in their Sunday finery at the Converse mill on a Sunday afternoon. Man in shirt on ground is Bill Fitzpatrick, later woods boss at the Shaver Lake Mills.

CHAPTER FIVE

THE HUMES FROM MICHIGAN

THOMAS HUME was a real-life Horatio Alger hero. By combining exceptional ability with hard work, he rose steadily from obscurity to multimillionaire in true story-book fashion.

Hume was born in 1848 at County Down, North Ireland, the first son and the second of an Irish farmer's eight children. The family was far from wealthy but the Hume ancestry contained a number of persons with unusual ability such as David Hume, the famous Scottish historian.

Young Thomas attended school until 14 years of age when he was indentured to a hardware and grocery firm, a normal procedure of the times. The apprenticeship paid him food and lodging but no money. Hume learned his trade quickly, received an early release from his indentures, and eventually advanced to company manager at an annual wage of $300.

In 1870, Hume set out for America in search of greater opportunity. Having relatives in Michigan, he soon found employment in that state as a tallyman and log scaler in the lumber industry. It was a cold and miserable job working winters in the woods as he did. In 1872, he was hired as a bookkeeper by the Muskegon, Michigan lumber firm of Hackley and McGordon. Hume mastered double-entry bookkeeping in a matter of several days! At the death of McGordon, who was killed in a saloon brawl in 1881, he was offered a chance to buy an interest in the Hackley organization and he seized the opportunity.

Hume promptly became a prominent factor in the business and was instrumental in the expansion of the firm's operations into diversified fields. He assumed full direction of the company in 1905 following Charles Hackley's death, and in later years was a leading officer in many widely separated activities ranging from pianos to telephones. Several times a millionaire, he achieved respect for his quiet but extensive philanthropy.

Hume was a stolid, cigar-smoking, church-going patriarch of the old school. His word was his bond and he took considerable pride in his unquestioned reputation for honest dealings.

Thomas Hume, a self-made man who achieved financial success beyond his fondest dreams. He was every inch the epitome of a "solid citizen." (Jean Hume Browning Collection)

Although a near genius in financial matters, he firmly believed that there was no substitute for hard work and raised his seven children (two sons and five daughters) in Victorian fashion.

Ira Bennett, the other protagonist in the new California lumber enterprise, was almost the very antithesis of Thomas Hume. A glib promoter, Bennett was always on the lookout for a sharp opportunity. His spendthrift ways kept him in continual financial hot water.

Bennett began his unlikely association with Thomas Hume in 1889 when the Bennett Lumber Company of Muskegon, Michigan, a yard-type operation owned jointly by Ira and his brother Jay, purchased lumber from the wholesale dock of Hackley and Hume. A few years later, the Michigan timber gave out and the Bennetts moved their company to Sandusky, Ohio. Hackley and Hume were directors of the Bennett organization and invested $40,000 — a considerable sum at the time — in the firm. Ira's main duties seem to have been that of salesman, as he traveled extensively throughout the middle west. He soon tired of the association and drifted into other lumbering activities after assigning his company stock to brother Jay under somewhat questionable circumstances. The Bennett Brothers operation later went into receivership and Jay moved west to Seattle, Washington. Ira turned up as manager of the American Lumber Company of Albuquerque, New Mexico, a position he held when he began negotiations to purchase The Sanger Lumber Company, as described in the previous chapter.

It was not at all surprising that Bennett turned to his wealthy, one-time associate, Thomas Hume, in an effort to obtain financial backing. Hume, like many eastern and middle western lumber barons, had been steadily acquiring virgin timberland in other states and Canada following the exhaustion of Michigan timber in the mid 1890's, and a move into California at this time was a natural step. Bennett's proposition interested Hume despite the rather unhappy result of their previous relationship.

But the Michigan lumberman was not the type to act without first obtaining a thorough knowledge of the facts. "When in doubt, it is better to do nothing," he often said.

Ira Bennett (Author's Collection)

In the summer of 1905, Hume commissioned J. D. Lacey and Co., a nationally known timber factor, to investigate the Sanger property and provide a timber cruise report. In addition, George Hefferan, a Michigan business associate, and George A. Hume, Thomas Hume's eldest son, were sent to California to personally inspect the operation.

It was obvious to all concerned that the Converse Basin was nearly cut over, but other great stands of redwood, fir, cedar and pine still remained on the 20,000 acres involved in the proposed deal. (The Sanger Lumber Company had previously sold some of its original holdings north of the Kings River to the Fresno Flume and Irrigation Company, a neighboring lumber enterprise.) Reports indicated that the most favorable future logging potential was centered several miles east of the basin and that a new mill in that area would probably be required. Even considering that prospective expense, the outlook seemed promising. Influenced by Bennett's glowing description of the country, and encouraged by a favorable timber cruise (in excess of 500,000,000 board feet), Hume decided to proceed with the purchase late in 1905.

Toward this end, a Michigan corporation, the Hume-Bennett Lumber Company, was formed on November 22, 1905, listing Thomas Hume, President; Ira Bennett, vice-president; George Hefferan, treasurer; and George A. Hume, secretary. Of the 2000 shares of preferred stock, Thomas Hume owned 1850 shares and the remaining 150 shares were divided equally among the three other participants. A total of 2000 shares of common stock was also issued.

The new corporation assumed the assets of The Sanger Lumber Company in December, 1905, for a price of $386,000 made up as follows: $1000 option received from Ira Bennett (later credited to his stock); six notes at $50,000 each, payable in three years; and a down payment of $85,000 consisting of $25,000 cash and a $60,000 credit allowed for the burning of the basin mill which had occurred while the sale was being discussed. The down payment was to cover the inventory which included all manufactured products and lumber, plus the merchandise at the company stores in Millwood and the basin. The inventory settlement was not finally resolved until October, 1906, when Thomas Hume voluntarily added $312.50 to The Sanger Lumber Company account "in the interest of harmony."

Original Hume-Bennett stock certificate issued to Thomas Hume in 1905. (Jean Hume Browning Collection)

Bennett's original option had called for a sale price of $405,000 which was apparently adjusted downward by negotiation. The fire damage to the mill was less than the $100,000 which had first been announced (the boilers, shafting and engines were virtually intact), with the result that the credit allowance because of the fire was set at the aforementioned $60,000.

Bennett, in his capacity as California agent and general manager, immediately set to work rebuilding the Converse facilities with the idea of continuing logging in the basin for a time. He employed W. R. Nisler, his assistant at Albuquerque, to supervise construction of a new mill. Bennett estimated that operations would begin by May 15, 1906, but it was July before the mill was actually cutting. The San Francisco earthquake and fire caused part of the delay because of the difficulty incurred in obtaining certain items of machinery and supplies from that battered city. Although plans called for only a temporary basin production and an early move to the east, the new double-band mill (capacity 175,000 board feet daily) was modern in all respects.

One reason for the decision to continue cutting at the basin was an excellent pocket of timber which lay nestled in a valley about a mile northwest of the Converse Hoist. To reach the area, an incline railway was constructed that dropped abruptly from the main tramway around a curve and into the valley. A landing point known as Camp Six paralleled the bottom of the incline.

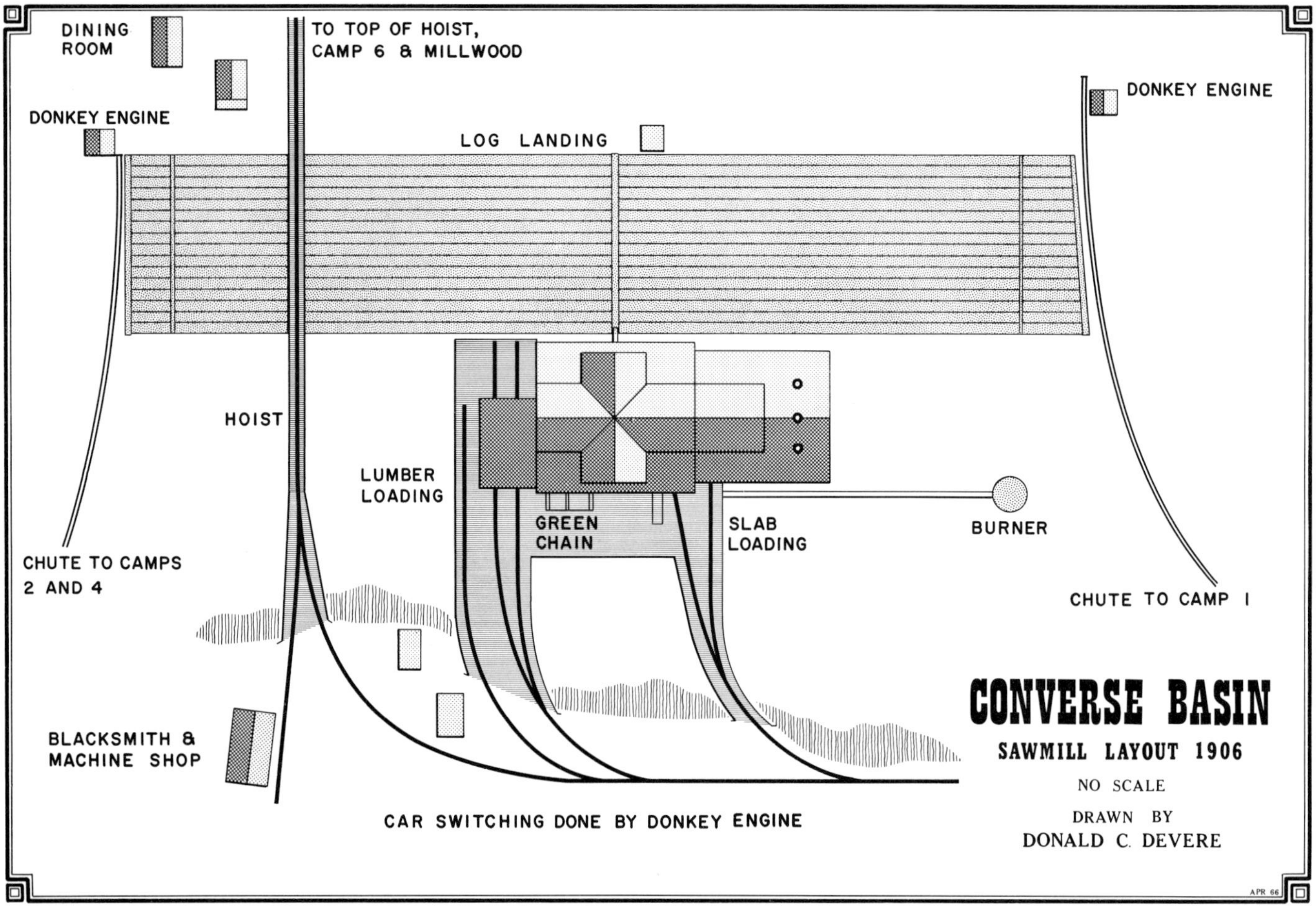

ABOVE: Sketch of the second Converse mill showing the log and lumber loading arrangements in detail. (Don DeVere)

BELOW: View of rear of second mill. Note that tramway does not go through building as in first mill.

Logs were brought by chute to the landing and loaded on railroad cars which were then pulled by heavy cable up to the hoist summit. At the top, the cars were switched onto the main tramway and lowered down to the basin mill. Handling the big logs on the ingenious double-incline system proved to be a delicate and nerve-wracking procedure. Caution was the watchword for the hoist engineers as the slightest miscalculation might bring a shower of ponderous timber deluging down on the unlucky crews working below.

But the loggers were adept at their trade and the heavy loads were pulled up one side of the mountain and lowered down the other in steady succession for several seasons without serious difficulty.

Cutting was also conducted by the new company at the small Abbott Mill (Camp Three) and logs were supplied for this purpose from a chute camp (Camp Seven) a short distance away.

Total production for the year 1907, the first full season of operation, was less than 20,000,000 board feet. In an effort to keep the Sanger finishing mill busy, Bennett bought additional lumber from one William McKenzie who had two small mills near Millwood. Boards were lowered by railcar down a rickety incline from the southernmost McKenzie mill to Millwood and then flumed to Sanger in the same manner as the Converse lumber. This arrangement unexpectedly terminated when McKenzie, who was heavily in debt, suddenly departed the country just ahead of his creditors. It was soon discovered that much of the lumber purchased by Bennett from the bankrupt McKenzie came from timber that had illegally been cut on Hume-Bennett's own lands! The trespass caused the company considerable expense and Ira Bennett no little personal embarrassment.

By the end of 1907, it was apparent that the main mill would have to be moved at the earliest opportunity if future cutting were to be profitable. The Hume-Bennett books showed a purported net gain of $53,136 for 1906 and $112,937 for 1907 but most of the usual expenses had been charged off to construction in the interest of a favorable depreciation account. In view of subsequent financial losses, it is much more likely that the two-year Converse operation proved extremely costly to the company.

Early in 1908, the decision was made to leave the basin. The Muskegon stockholders, who had kept themselves well informed of the situation,

McKenzie tramway from upper McKenzie mill to Millwood. Lumber was shipped by car to Millwood for transporting in the flume to Sanger. (Dick Challacombe Collection)

were responsible for major policy changes such as this, although Ira Bennett continued to handle the local direction of the company. George Hume and George Hefferan paid several visits to California in both 1906 and 1907, and Thomas Hume himself journeyed to Sanger while on a western tour in October, 1906.

The site chosen for the new operation was Long Meadow, a beautiful, grassy tableland some four miles due east of Converse Basin at the confluence of Tenmile and Long Meadow Creeks. Preparations were made to erect a large dam to stop the flow of the two creeks which drained a 25-square mile area, thus creating a permanent reservoir. The lake would provide needed log storage — lack of which had proved a serious handicap at the other mountain mills — as well

ABOVE: Log landing at foot of Camp Six incline. Logs were brought to incline by chute. (Dick Challacombe Collection)

BELOW: Big redwood logs at Converse mill landing. (Fresno County Historical Society Collection)

as a year-round water supply for the flume. Adjacent to the dam, an entirely new and modern sawmill was laid out that included two large kilns to expedite the drying of raw lumber before fluming. Contracts were also let for a 17-mile flume extension to connect the new mill with a portion of the existing flume so that lumber could be floated all the way to Sanger without interruption.

Building a major project of this sort in the rugged, isolated mountains that lay beyond Converse Basin presented a tremendously difficult and expensive challenge. The road up from the valley to Millwood was a narrow, tortuous affair with few turnouts, and there were no roads of any sort beyond Millwood.

After due consideration, two routes were chosen to reach the Long Meadow country which lay at a 5200-foot elevation. A crude road was cut that ran from Converse through Indian Basin and ultimately over a ridge into Hume — the name given to the new mill area. A second makeshift road took off from the existing Millwood Road, wound its laborious way through the hills and forests, and finally joined the first road well above Long Meadow. Neither road was much of a highway but both served the purpose of providing access for the multitude of equipment and supplies that were needed.

The Hume-Bennett Company was most fortunate to obtain the services of a remarkable and enterprising engineer to supervise construction of the new dam and mill. The man was John S. Eastwood, former engineer at the Fresno City Waterworks and a pioneer in the development of hydroelectric power in California. As the author pointed out in his book *The Railroad That Lighted Southern California* (Los Angeles: 1965), Eastwood originated the San Joaquin Electric Company, later to become the San Joaquin Light and Power Company, and was instrumental in the conception of the famous Big Creek hydroelectric complex of the Southern California Edison Company.

Eastwood designed a revolutionary multiple-arch dam, called by experts the first such reinforced concrete dam built in the United States. A multiple-arch dam consists essentially of a series of arches carried by buttresses spaced at suitable intervals. The upstream faces of the buttresses usually have considerable slope, and the distribution of foundation pressure is largely controlled by the water weight in the arch extra-dosal areas. As a result, the yardage of concrete is substantially less than that required for a solid-gravity dam. This is of particular value where transportation of material is a major factor which was certainly the case in the Hume project. In more recent years, many similar arch dams have been constructed, but at the time, Eastwood's idea was a real innovation and attracted considerable notice in engineering circles. Thomas and George Hume, always progressive in their business ideas, agreed to accept the radical new design.

John Eastwood, one of the ablest and most inventive early engineers. He supervised the construction of Hume Dam.

In the early spring of 1908, as soon as weather permitted, teams began hauling freight and machinery up the rough road from Sanger to Hume. It took a skillful driver and a well-trained team of 12 or 14 mules to negotiate the tricky turns and the narrow clearances involved in the mountain route. Much of the raw material for the dam was sent directly to Millwood, loaded on railroad cars and taken up the Converse Hoist. From this point, it was carried overland to Hume by wagon.

View of the big mill at Hume while construction work was under way. This picture taken April 11, 1908.

Work on the mill itself began in March, 1908. A small temporary mill was first set up to cut the necessary lumber required in the permanent structure. At the same time, clearing proceeded on the dam site. By June, crews were blasting out the dam foundations, and within a month, a construction trestle was in use across the gorge. The first concrete was poured into the dam forms on August 15, after a 24-day delay because of lack of cement. In the meantime, a large force of laborers was removing trees and debris from the reservoir site.

It was a hectic and noisy summer at Long Meadow in 1908. The stacatto chatter of carpenters' hammers and the cacophonous pounding of heavy concrete mixers created a strange discordance in the solemn and lonely Sierra. In surroundings which resembled a travel folder photograph, busy crews found little time to savor the scenery as the job was carried on from dawn to dusk, seven days a week, in a frantic attempt to complete the complex program before winter thwarted further efforts.

Finally, in November, even as the first snow began to fall, the dam was finished. The reservoir stood empty until the following spring so the dam could be thoroughly tested for cracks. When all appeared in order, the arched portals that had been left open so winter freshlets could pass, were closed, and in June, 1909, the 87-acre lake began to fill. The reservoir had a capacity of 1411 acre feet of water and was 50-feet deep at the deepest point.

Three outlets were provided in the dam: one (24-inch diameter) for drawing off water at low stage to clear the pond of logs, and two (12-

The Hume Dam when partially finished. Note temporary flume in foreground and arch work construction.

inch diameter) for the discharge of water into the flume. The flume openings were screened with railroad bars set in concrete boxes with a flap-gate of wood provided in case the regular valves needed emergency repairing.

The Hume Dam was 677-feet long on the center line of the crest, 61-feet high at the highest point, and contained 2201 cubic yards of solid concrete plus eight miles of old steel cable. It took 114 days in all to build the dam, and the total cost was $46,541. Eastwood estimated that a regular, filled dam of like capacity would have cost $74,000. It is interesting to note that laborers were paid $2.25 per day, skilled labor $3.00 per day and carpenters $3.50 per day on the Hume project. Cement averaged $5.00 per barrel delivered at the site from Sanger, 58 miles away via rough roads, railroad cars and hoist.

Eastwood's arch-type dam was so unusual that he received inquiries from many foreign countries concerning the idea. He later was the design engineer for 16 other dams of like construction and achieved a measure of international fame as a result. His untimely death by drowning in the Kings River on August 10, 1924 cut short his brilliant career.

* * *

The moving of the railroad and mill equipment from Converse Basin posed a tremendous problem for the Hume-Bennett crews because of the immense weight and unusual size of the items involved. A special wagon was constructed with four wheels in front, much like a modern truck, and 12 mules were required to haul the heavy load. Most of the railroad iron was pulled up and transported to Hume for relaying. The bulky

ABOVE: Panoramic shot of mill area showing railroad trestle at left, mill buildings, flume and dry-kiln tramway. (Annie Perry Collection). BELOW: Building the railroad grade east of Hume. This was a deep rock cut and the crews were hard pressed to prepare a proper road bed because the equipment was of the most rudimentary sort. About 15 miles of track eventually were constructed. (Jack Clark Collection)

Shay locomotive was shipped in sections by wagon, a backbreaking job in itself.

When all items of worth had been removed from Converse, the basin mill, no longer of any logging value, was deliberately set afire and allowed to burn to the ground. This somewhat ignominious conclusion to Converse Basin lumbering marked the end of perhaps the most lurid chapter in the history of American logging.

By the spring of 1909, the Hume mill was cutting and crews were hard at work laying railroad track into the woods along the east side of the lake. Another work force, under special contract, had begun the 17-mile flume extension which started from the dam and headed down Tenmile Creek on its way to the Kings River Canyon, a drop of about 2500 feet in elevation. At the river, the flume turned west and paralleled the stream, although high above it in most places, to the eventual junction with the existing Millwood flume. The new flume passed through territory far more rugged than anything previously encountered by flume builders. Sheer granite walls, steep rocky gorges, rattlesnake-infested ravines and jungle-like underbrush confronted the intrepid workmen as they battled foot by foot, to raise the

spidery trestles and install the long flume boxes that were required so at least the semblance of a proper grade could be maintained. In some places, horizontal foundation beams were driven directly into the cliff sides and anchored by steel cables hung from above, as there was not sufficient room for the usual upright supports. Several workers lost their lives at one point when a high trestle upon which they were climbing suddenly collapsed after being overburdened by a heavy load of supplies arriving unexpectedly by flume boat. As the shattered structure toppled into the canyon far below, the helpless men were dashed against the rocks and killed.

The tedious and daring job was finally completed in the fall of 1909 when connections were made with the old flume at the point where it turned south from the river on its way up to Millwood. On September 29, 1909, water began flowing into the trough from Hume Lake.

The 17-mile extension cost more than $100,000 to construct or about $1.00 per running foot! Total length of the extended flume was 59 miles, making it probably the longest lumber flume ever built in the world. There were 17 flume stations on the new route — 13 equipped with telephones

BELOW: This picture of Millwood taken shortly after the company moved operations to Hume illustrates the quick collapse of the former mountain resort and lumber camp. Within only a few years, many of the buildings were burned or torn down, and the rest abandoned. Today, there is not a single obvious trace of the once prosperous and thriving town that served as the flume origin point.

THE HUME FLUME

These remarkable pictures of the building of the 17-mile flume addition from Hume to the junction with the old flume belong to John Cummings, whose father was a member of the construction crew. The country was unbelievably rugged and the job was a monumental one. Note the flume boat (below) which is loaded with supplies as it crosses one of the highest trestles on the route. (All photos John Cummings Collection)

— a lengthy and highly unusual party line!

Unlike the upper Millwood flume, the new section from Hume to the Kings River was NOT recommended for flume boats. Only the most audacious riders dared to venture down the wicked Tenmile Creek section on one of the V-shaped rafts. The route traversed a seemingly endless succession of sharp curves, high trestles (one was 107 feet tall) and breathtaking descents. At the worst point, the grade was so steep that a boat, rushing downward at a tremendous pace, would push all the water before it. At the bottom of the abrupt fall, the rider could look back and see the flume dry for some distance. All at once the water would come rushing up to the boat again and suddenly, with a great bumping and splashing, the clumsy craft was once more in rapid motion along the perilous course to the river far below, the frantic rider hanging on for his very life.

Several persons were severely injured in separate accidents while attempting to ride the Hume flume. Even those who made the trip unscathed were loathe to try it a second time.

"I was attracted to that doggone flume like some people are to rattlesnakes," recalls one old-time Hume lumberjack. "It was the quickest way to the river when I wanted to go fishing, and I'd get up just enough nerve to set out a boat. Then I'd take one look down that crazy, winding devil of a route and decide that it was better to walk and arrive late than to push off on that boat and maybe not arrive at all. At least not in one piece!"

* * *

The new Hume operation did not really begin full production until the summer of 1910. Although some 8,000,000 board feet of lumber were cut at the mill in 1909, construction and moving activities occupied the major portion of the season.

The Hume Lake area resembled a wild-west camp in many ways. The sawmill stood immediately adjacent to the dam. Built at a cost of $150, 000, it was a sturdy, three-story structure, 50 feet wide by 133 feet long, set on concrete piers. As with most early mills, the ground floor contained the heavy machinery; sawing was conducted on the second floor. Logs were dumped into the lake from a railroad landing above the opposite side of the dam, and stored in the water until needed. A little car, set on wide-gauge railroad tracks, ran up a ramp from the lake and entered the mill at the second-floor level. On its downward run, the carriage car dropped below the water

Workmen heading down to the construction point while building the Hume flume. (John Cummings Collection)

surface so a log could be easily floated into proper position atop the car. At a signal, the hoist operator reeled in the cable and pulled the car with its load into the mill.

The log was received on what was called a log deck. From there it was rolled on the carriage with a log turner, sometimes called a "nigger." The carriage then propelled the log into the big band saw — a narrow ribbon of endless steel running at great speed over two pulleys. The first sawing consisted of cutting slabs from two sides of the log; then the carriage might be returned and each board cut off, one at a time, or the trimmed log might be sent on to the smaller band saw depending on the size of the timber in question.

The main floor also held the edger, rollers and gang circular saws which sliced out many pieces of the proper-sized timber with one forward movement. The saw-filing room was in the third floor loft.

From the mill, the boards were conveyed over a trestle into two huge dry kilns which could accommodate 2,000,000 board feet of lumber at one time. After only three days in the kilns, the raw lumber was ready for clamping and fluming. At Millwood, the seasoning process had

TOP: Hume store and post office. CENTER: Dry kilns near mill where lumber was dried before fluming. BELOW: The "piano player" inside the mill who selected the proper saw combination for maximum lumber cut. (All photos Jean Hume Browning Collection)

taken several weeks or more depending on the weather and the stacking procedure.

The big mill was largely steam powered although a small generator provided enough electricity for a rudimentary lighting system. The concrete boiler house, 50 by 78 feet, and engine house, 26 by 56 feet, were alongside the main building. A cluster of other structures included a pump house, foundry, blacksmith shop, locomotive engine house, machine shop, cook house, combination hospital and barber shop, and later, a schoolhouse and a dance hall. The main warehouse and office, 17 by 65 feet, also served as a store and post office. About 30 houses for the married men and a group of 100 cabins for the single workers provided the principal accommodations for employees. Except for the concrete engine and boiler houses, all buildings were wood — unpainted on the outside and unplastered on the inside. Kerosene lamps, outside toilets and windows draped with mosquito netting contributed to the rustic appearance of the camp.

Framed by the surrounding mountains, Hume Lake seemed picturesque enough to have been a remote movie location, which indeed it was on several later occasions. The bucolic atmosphere was shattered only by the pounding of the heavy mill equipment, shrill blasts from the big steam whistles and the occasional clatter of the wood-burning locomotive, *Sequoia,* as it forged its laborious way to and from the timber. The narrow gauge railroad originated near the mill, ran across the top of the dam, and then wended its way into the woods on the east side of the lake. Logs were pulled to landings along the line by donkey engines situated near the cutting areas. After being loaded, the log cars were hauled back to the log-dump trestle along the lake, and their cargo deposited into the water by means of an A-frame crane.

The trusty little Shay, only motive power on the line at this time, performed solitary and yeoman service during the first two seasons at Hume in a Herculean effort to supply the mill with sufficient timber. Fortunately, most of the early logging was in the immediate vicinity of the lake and the time involved in running to and from the cutting areas was not great.

Besides the great mountain construction program, improvements and additions were made concomitantly in the rest of the lumber operation. The Sanger finishing mill and yards were updated and expanded. A new wood-and-steel flume bridge replaced the decrepit suspension structure

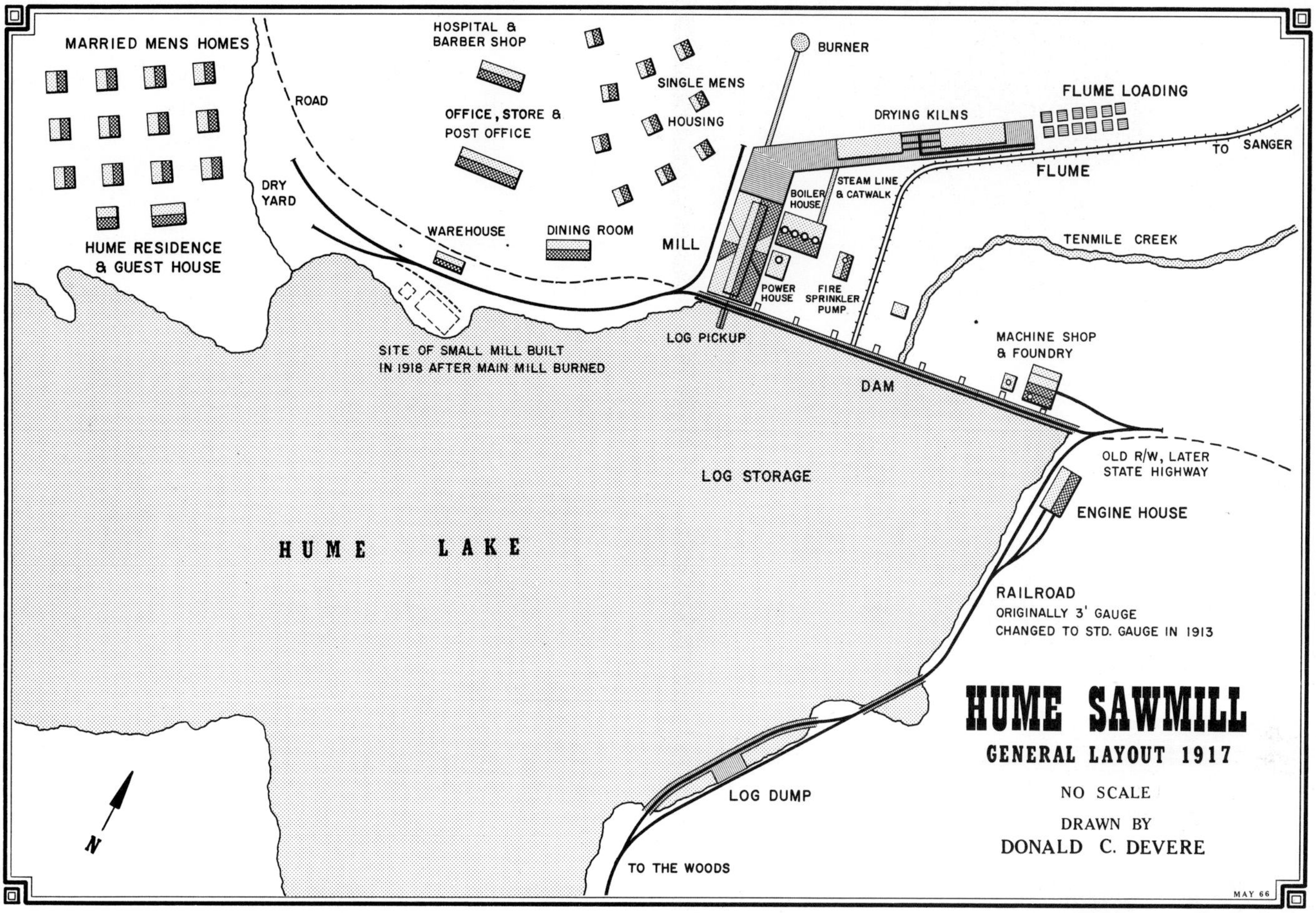
MARRIED MENS HOMES
HOSPITAL & BARBER SHOP
BURNER
ROAD
OFFICE, STORE & POST OFFICE
SINGLE MENS HOUSING
FLUME LOADING
DRYING KILNS
TO SANGER
FLUME
DRY YARD
STEAM LINE & CATWALK
BOILER HOUSE
WAREHOUSE
DINING ROOM
MILL
TENMILE CREEK
HUME RESIDENCE & GUEST HOUSE
POWER HOUSE
FIRE SPRINKLER PUMP
LOG PICKUP
SITE OF SMALL MILL BUILT IN 1918 AFTER MAIN MILL BURNED
MACHINE SHOP & FOUNDRY
DAM
OLD R/W, LATER STATE HIGHWAY
LOG STORAGE
ENGINE HOUSE
HUME LAKE
RAILROAD
ORIGINALLY 3' GAUGE
CHANGED TO STD. GAUGE IN 1913
HUME SAWMILL
GENERAL LAYOUT 1917
NO SCALE
DRAWN BY
DONALD C. DEVERE
LOG DUMP
N
TO THE WOODS
MAY 66

16 DAM&MILL HUME CALIF.

The Hume-Bennett finishing mill and box factory at Sanger about 1912. (Jean Hume Browning Collection)

across the Kings River. By the end of 1910, Hume-Bennett Lumber represented an investment of $1,600,000 and ranked among the finest and most modern logging plants in the entire Sierra.

Despite the excellent physical condition of the company, Thomas Hume was not at all satisfied with the financial progress of his California enterprise. For some little time, in fact, he had felt that the heavy expenses incurred by the company were badly out of line with the limited results. His displeasure began to focus more and more on the actions of the general manager, Ira Bennett. Even before the costly move to Hume Lake in 1908, the relationship between the two lumbermen was rapidly deteriorating.

In 1905, when Bennett first approached Thomas Hume regarding The Sanger Lumber Company purchase, the Muskegon businessman assumed a cordial and somewhat paternal attitude toward his younger associate. He was undoubtedly impressed by the skillful sales presentation; possibly he was sympathetically drawn to Bennett because of the untimely death of the latter's young son from typhoid about this time. Whatever the case, Hume apparently had some confidence in Bennett's ability; enough, at least, to go ahead with the expensive California lumber investment.

By early 1907, however, after less than two seasons of logging at Converse Basin, the feeling between the two men had undergone an unexpected and drastic metamorphosis.

"Your woods cost figures are excessively high and your general expenses seem exorbitant," Hume charged in a 1907 letter to Bennett. "Your trial balance shows frightful costs, and sawing reports show very little results for the expense. Are you using some of the money for your own needs?"

Hume's questioning attitude regarding the high California expenses was caused in part by Bennett's personal financial adventures, which were unusual to say the least. In addition to his company duties, Bennett was a sometime dealer in High Sierra timberlands. Apparently the sideline was not too successful, for he seemed to sink deeper into debt with every manipulation.

One of Bennett's timber deals proved of particular annoyance to Hume. In 1907, the Michigan lumberman was negotiating to buy as a separate investment, a magnificent stand of timber situated about 45 miles south of Hume Lake. The property, later known as the Mountain Home Tract, consisted of a total of 4562 acres — many covered with huge redwoods — and was a collection of the individual properties of eight separate owners. Bennett, who had an interest in a small portion of the land in question, became involved in the transaction at an early date and made it known that he expected to net a "handsome profit" on the sale. Hume felt that under the circumstances, Bennett was not ethically entitled to any profit at all, and he quickly instructed his Fresno attorney to keep all knowledge of future purchases in the tract away from Bennett. The affair was eventually resolved by giving Bennett a credit against some of his company debts in exchange for his Mountain Home equity.

In August, 1909, Bennett reported that he owed $37,000 and that his only assets were his home and his Hume-Bennett stock. In November, 1909, he borrowed $50,000 from the firm of Livingston and Hovey of Detroit, Michigan, pledging his stock as collateral. Within a short time, he attempted to arrange a loan from Thomas Hume but was refused.

Bennett was cautioned to "stop excessive spending" and "correct your constant misstatements" on several occasions. His car allowance was discontinued and he was warned repeatedly about overdrafts on his own personal salary account. On December 2, 1909, the unpleasant situation culminated when the Hume-Bennett directors voted (Bennett objecting) to remove Bennett from his position as manager of the company and took away his authority to purchase or incur expense. He was replaced by T. W. Decker, a knowledgeable lumberman, in the hope that some financial recovery could be effected.

There was unfortunately no immediate improvement.

"You received $45,000 this month and are in just as bad shape as before we sent it," the Muskegon office noted in a directive to Bennett and Decker on August 30, 1911. "You have had $120,000 since January 1, and from all we can see now we are not finished yet if we want to keep the operation going."

Besides subsidizing the company, Thomas Hume felt compelled to keep up the interest on some of Bennett's personal loans since they were guaranteed by Hume-Bennett stock.

Finally, on January 22, 1912, Bennett was forced out of the California logging operation. He resigned all rights and relinquished his stock and other interests to the remaining stockholders who met the various obligations outstanding against the certificates.

Undaunted by his ouster, Bennett immediately put his promotional abilities to work in another direction and on December 5, 1912, less than a year after his departure from the Hume-Bennett Company, negotiated the purchase of the Fresno Flume and Lumber Company* of Clovis, California. He was financed in the deal by Clarke L. Poole and Company, a Chicago investment firm.

The electric plant at Hume which provided some interior mill lighting. Jack Clark, later mill superintendent, at left. (Jack Clark Collection)

The acquisition was particularly annoying to the Humes because the nearby Clovis enterprise (10 miles north of Sanger) was a major competitor with an operation very similar to that of Hume-Bennett.

After only one unprofitable year, the Poole Company abruptly withdrew its backing. Amazingly, Bennett was somehow able to persuade another Michigan lumberman, Horatio Hovey, to take over the financing and continue the enterprise. The new arrangement also lasted only a single lackluster season and the firm became inactive in 1915. Four years later, the property was sold by the bond holders to the Southern California Edison Company and became a part of that corporation's Big Creek hydroelectric complex.

Bennett left the area and entered the mining business near Prescott, Arizona. He was never again a factor in the Sierra logging industry.

Through all of Bennett's machinations, Thomas Hume kept scrupulously silent about his unsatisfactory and costly association with the profligate promoter. He knew he had erred in becoming involved in the California lumber deal, but he was the sort of man who faced his obligations, whatever the cost, and made the best of the situation.

Hume's feelings about Bennett can be determined very well, however, by a remark he made some years later when speaking to a friend about a man who wanted to sell him a gold brick:

"It made me wonder if his first name wasn't Ira B."

* * *

At the time of Bennett's ouster in 1912, the Hume-Bennett Lumber Company was like a mill-

*Formerly the Fresno Flume and Irrigation Company. The corporate name was changed in 1908. The firm had a sawmill and pond at Shaver Lake, and a 42-mile flume running to Clovis.

A break in the Millwood flume about seven miles below Millwood. The structure had settled on a curve and lumber washed down the hillside. (Harold G. Schutt Collection)

stone around Thomas Hume's elderly neck. The Sierra enterprise represented an investment of more than $2,000,000 and was losing money at an unrelenting pace. If he could have done so at a reasonable price, Hume would surely have sold the far-west logging show without regret. He must have wondered how he had ever become involved in such a troublesome venture some 2500 miles from home.

At this point in his life, Hume was too old and too busy to take any sort of active role in the distant California operation. Moreover, his biggest philanthropy, the Old People's Home (now called The Hume Home), was in the process of being built in Muskegon and occupied much of his time and interest. Yet the costly Hume-Bennett Lumber Company was like a burr under a saddle blanket. It required thorough and immediate attention, and in particular, a superintendent of capability who could assume direct control of the local management.

Finally, Thomas Hume turned to the one available man in whom he had sufficient confidence and trust — his eldest son, George Alexander Hume.

When he moved west in 1912, George Hume was not quite 31 years of age and he looked more like 20. Completely unlike his straitlaced father in most ways, the young man from Michigan was an eccentric and refreshing personality who burst upon the tranquil Sierra scene like a thunderbolt! Some of his unusual behavior might have been the result of his sudden release from direct and constant control, as Thomas Hume was a "spare the rod — spoil the child" autocrat who used the hickory stick to good advantage. Even after he was married in 1904, George was expected to

LEFT: The old flume bridge across the Kings River stands beside the new bridge built in 1910. NEXT PAGE TOP: Dumping logs at the Hume log dump. The device behind the splashing water is a "mallard" which was used to raise sinkers from the lake. BOTTOM: The One Spot pushes two cars on to the dumping trestle preparatory to unloading.

LOGGING TRAIN HUME CALIF.

Head of the flume at Hume. Boards came down the ramps and were clamped at bottom right. (Jack Clark Collection)

have the spittoons cleaned and the Muskegon office swept out and open for business by 7 o'clock in the morning.

In many ways, George Hume was an enigma. It would have been hard to find a millhand who looked less like a millionaire. He delighted in playing the lumberjack and often went to town sporting spiked boots, stagged pants with a rope for a belt, and a three-day growth of beard. When the local hotel clerks were about to eject him, he would suddenly produce a huge roll of bills from his pocket, and with a flourish, settle his account in advance.

He drank too much, smoked too much (three packs of cigarettes daily) and seemed to take an almost fiendish satisfaction in shocking the natives whenever possible. There were rumors that George had an overdeveloped eye for a pretty girl, but his capable wife Nan remained resolutely unperturbed by all such hearsay. Privately, his closest friends confided that he actually encouraged gossip of this sort to increase his reputation as a wild character.

Hume suffered from a number of allergies which proved extremely annoying. Eggs were like poison to him and milk was nearly as bad. His beloved English setters could not be kept in the house because their dandruff caused an unfavorable reaction. He was deathly afraid of water although he forced himself to swim on frequent occasions.

An outdoorsman by choice, he loved hunting and fishing, but he was also quite literate and could quote poetry by the hour. His father had not permitted college believing it a waste of time, and as a result, George attended business school and became very adept at mathematics, for which he had great natural ability. Strangely, his typing remained of the two-finger variety.

Through all his odd behavior and personal difficulties, young Hume somehow emerged as a tireless and proficient lumberman with a firm reputation for punctilious honesty. A 15-hour work day was not unusual for him, and his father often cautioned him about overdoing.

On many occasions George quietly aided hard-luck employees; yet he could be tough as nails when the circumstances demanded it. When he wanted to telephone Sanger on business, woe to the unlucky flume herders' wives who might be gossiping along the party line at that very moment. His stentorian order to "get the — — off the line" could be heard almost the entire length of the route without any aid at all from Alexander Graham Bell's remarkable invention.

George was followed to California by his wife and four children (a fifth and sixth were born later) and the family soon settled in a two-story house above Hume Lake. Winters were spent at Sanger or Fresno where full-time schooling was available. When the logging season was ready to begin each spring, Hume insisted on being the first person to reach the snowbound Sierra mill. It seemed a matter of personal pride for him to bring fresh fruits and vegetables to the several maintenance men who had stayed in the mountains all winter, and he sometimes hiked 30 miles or more up the flume with a heavy pack in order to achieve his purpose.

* * *

Hume's first efforts after taking over the company management were directed toward obtaining more proficiency in the woods and railroad operations. Working two shifts a day, the mill itself could handle far more timber than was currently being logged. Total cuts for 1910 (21,174,350 board feet) and 1911 (21,866,271 board feet) fell far short of the 35,000,000 board feet that were estimated as the break-even point.

A second Shay locomotive was purchased new from Lima in early 1911 after repeated attempts to buy a second-hand engine were unsuccessful. In 1912, Hume added a McGiffert Log Loader ($6500), seven sets of big logging wheels ($360 each) and a variety of lesser equipment including 40,000 flume clamps (8 cents each). The McGiffert Loader was a log-loading device supported on truck frames which rested on the outside portion of the railroad ties. It actually straddled the log cars. A swinging boom lifted logs from the ground — sometimes from several hundred feet or more away — and placed them on the cars as they passed underneath the loader. A complete log train of eight to 10 cars could be quickly prepared in this fashion. When ready to move on to a new point, the weight of the loader was lifted from the ties using cables which raised the truck frames and lowered railroad trucks to the track. The McGiffert was now self-propelled and could advance under its own power.

There were two rail lines in use. A common line ran south along Tenmile Creek from the log dump to a junction about a mile above the lake.

ABOVE: George Hume in a rare "bearded" pose. He wore the beard for only a short time and was usually clean-shaven. BELOW: The Hume home at Hume Lake. (Jean Hume Browning Collection)

ABOVE: Demonstrating the McGiffert Loader after its arrival in 1912. (Jean Hume Browning Collection). BELOW: A head-on look at the McGiffert illustrates how the contrivance could travel on its own wheels.

Here one line headed back to the north on a switchback and then east across Tornado Creek toward the redwoods. The second railroad, called the pine line, continued south up Tenmile Creek to Bearskin Meadow. Several spurs reached into timber pockets on either side of this line (see map) and heavy logging of fir, sugar pine and yellow pine was conducted. At Camp Five, near Bearskin Meadow, big wheels — 10 to 12 feet in diameter — were used to bring logs to the railroad where the McGiffert Loader placed them on cars.

The big wheels were rolled over downed timber in the woods and a tongue attached to the front of the log with chains. Four-horse teams then pulled the tongue to a horizontal position which raised the front end of the log from the ground and enabled it to be dragged with relative ease to the loading point. Hauling of this sort was possible only if the terrain were approximately smooth and the distance short. It was not particularly efficient under even ideal conditions.

On the pine line, between Camp Three and Camp Five, a spur line ran toward Lava Butte where a 1500-foot incline with a 30 per cent grade reached into the woods near a point called Finger Rock (now known as Buck Rock). A huge Tacoma bull donkey situated at the head of the incline pow-

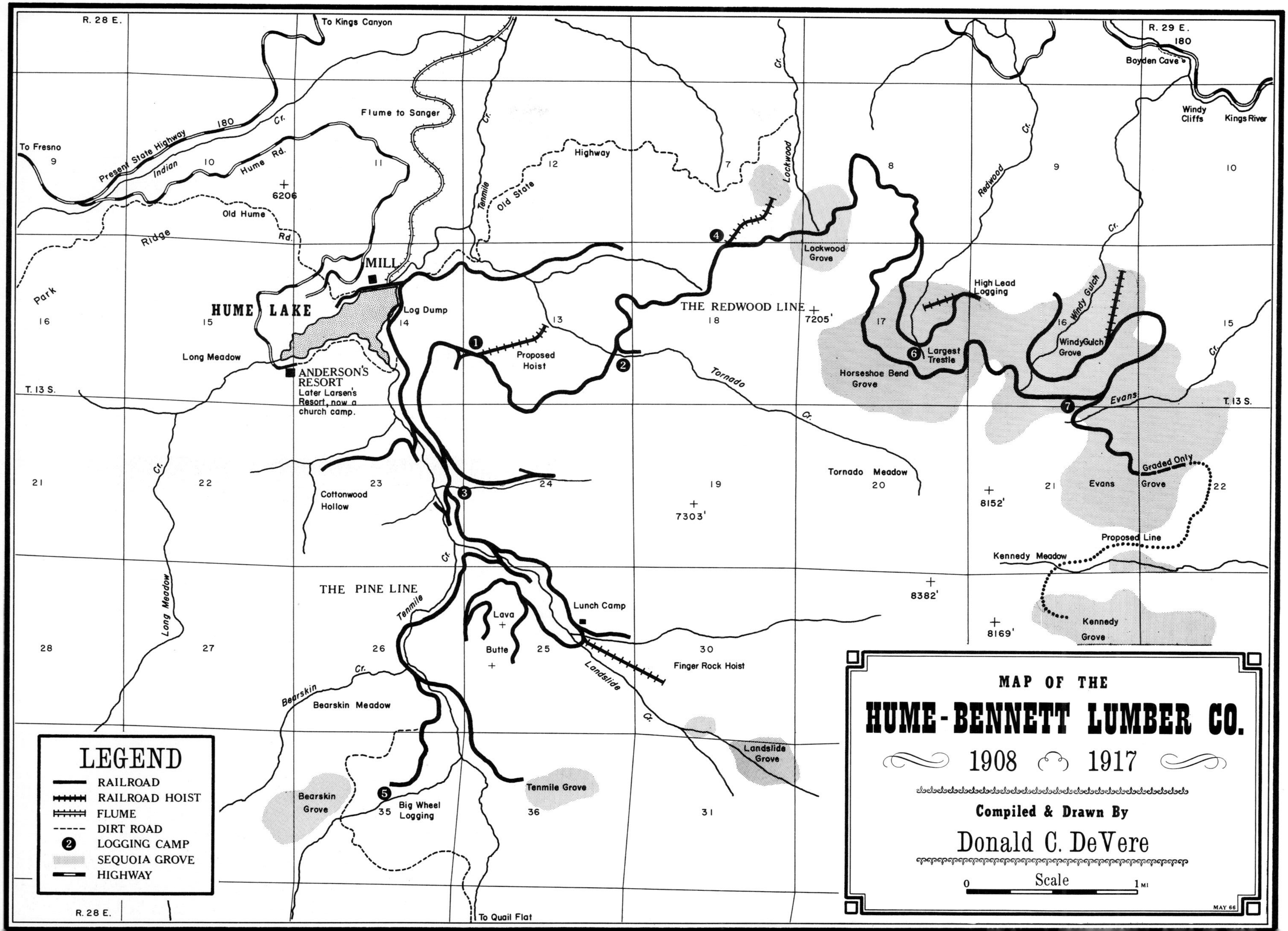
MAP OF THE
HUME-BENNETT LUMBER CO.
1908 1917
Compiled & Drawn By
Donald C. DeVere
Scale
0
1 MI
MAY 66
LEGEND
RAILROAD
RAILROAD HOIST
FLUME
DIRT ROAD
LOGGING CAMP
SEQUOIA GROVE
HIGHWAY
R. 28 E.
R. 29 E.
T. 13 S.
To Kings Canyon
To Fresno
To Quail Flat
Present State Highway
180
Indian
Cr.
Hume Rd.
Old Hume
Rd.
6206
Flume to Sanger
Tenmile
Old State
Highway
Ridge
Park
MILL
HUME LAKE
Log Dump
Long Meadow
ANDERSON'S RESORT
Later Larsen's Resort, now a church camp.
Proposed Hoist
THE REDWOOD LINE
7205'
Lockwood
Lockwood Grove
High Lead Logging
Largest Trestle
Horseshoe Bend Grove
Redwood
Windy Gulch
WindyGulch Grove
Evans
Tornado
Tornado Meadow
Boyden Cave
Windy Cliffs
Kings River
Graded Only
Evans Grove
8152'
7303'
Proposed Line
Kennedy Meadow
8382'
8169'
Kennedy Grove
Cottonwood Hollow
THE PINE LINE
Lunch Camp
Lava
Butte
Finger Rock Hoist
Landslide
Landslide Grove
Bearskin
Bearskin Meadow
Bearskin Grove
Big Wheel Logging
Tenmile Grove

ABOVE: Big Wheels were used near Camp Five. (Jean Hume Browning Collection)

BELOW: Big wheels bring logs to the railroad for loading using the McGiffert Loader. (Harold G. Schutt Collection)

ered the heavy steel cable which let loaded cars down and pulled empties up, two cars at a time. On more than one occasion, the cars broke loose from the cable or from each other and sailed unattended down the steep grade. In the event of such a runaway, the hoist engineer quickly sounded the alarm with a frantic and shrill blast of his steam whistle. The loading crews at the bottom of the hill, at once aware of the danger, scattered for safety with a high degree of dispatch.

In 1912, as a result of all the improvements, a record 27,003,873 board feet of lumber was flumed to Sanger from the Sierra operation. Nearly 1000 men were employed in all and the mill itself was running two 10-hour shifts. For the first time in years, the Humes had some cause for optimism about the enterprise. George was even seriously contemplating the moving of the Sanger yards to a place above the Kings River flume crossing, which would considerably lessen the time required in flume transit.

Besides the heavy logging activity, the Hume Lake area was booming in other ways. Tourists in ever-increasing numbers were finding their way to the rugged and beautiful Sierra location. Campers, hikers, fishermen and nature lovers of all sorts came from the hot San Joaquin Valley to the cool and restful sylvan wonderland for relief and relaxation, although their presence was not always welcomed by the busy logging crews. A new road was under construction by the government, and motorized trucks were about to begin regular service from Sanger.

But just at the very time when prosperity appeared to be most imminent, the bottom dropped out of the lumber market with a resounding thud. The nation as a whole suddenly plunged into a three-year period of recession and the construction industry, as usual, was among the first and hardest hit. As a result of the lack of business, Hume-Bennett production nose dived to 15,811,120 board meet in 1913 and even worse, to 11,382,480 board feet in 1914. The poor showing was a great disappointment to George Hume, who had worked tirelessly, and very costly to Thomas Hume, as the latter was forced to donate $800,000 in cash to the company in 1913 in order to keep it operating.

"Even money seems to appreciate the California climate," the elder Hume wryly observed. "It refuses to come home."

Efforts were renewed to sell the enterprise. "Work toward the point of getting out of Fresno

ABOVE: George Hume, complete with rifle and supplies, poses on a flume boat somewhere along the lengthy route. BELOW: The wagon road into Hume was no super highway in the early days! (Both Jean Hume Browning Collection)

County altogether," Thomas Hume told George in August, 1914. "I simply hope that by operating in a limited way for a limited time, we may find some way of getting out and getting at least part of our money back."

Elmer Cox, general manager of the neighboring Madera Sugar Pine Company, evinced a mild interest in the purchase of the Hume operation. Some years before, he had also considered buying the old Sanger Lumber Company, as mentioned earlier. Cox came to Hume Lake in 1914, looked over the equipment, and made what seemed like a ridiculously low offer for the company. Thomas Hume refused.

The idea was advanced by George Hume of combining the four neighboring Sierra logging companies* into one large corporation, but the unwelcome presence of Ira Bennett at the Fresno Flume and Lumber Company precluded any serious moves in this regard.

"I still believe that within a few years the lumber business will take a turn so that we can make some money with that plant out there in California," Thomas Hume observed. He determined to make the best of the bad situation and wait for an improvement in the general economy.

It appeared that the redwood market would be more active than the pine market in the immediate future. Moreover, the competition was less keen in the redwood field at this time. Accordingly, at the start of the 1914 season, the entire railroad was converted from the previous narrow gauge

*Hume-Bennett Lumber Company; Yosemite Lumber Company, Merced; Madera Sugar Pine Company, Madera, and Fresno Flume and Lumber Company, Clovis.

This amazing panoramic view was taken July 8, 1911 and gives an excellent picture of log-loading methods in the Hume operation. A "cherry picker" at left swings logs onto the narrow gauge cars at the landing. Yarding donkeys at right were used to drag timbers to the landing from the cutting areas. (Jean Hume Browning Collection)

(3 foot) to the more usual standard gauge (4 foot 8½ inches) with the intention of continuing the eastern rail line into the big trees. The narrow-gauge cars were too small and too light to transport heavy redwood loads for any distance, and the closest redwood groves were more than six miles by railroad from the lake. The two Shays and the McGiffert Loader were rebuilt to the new gauge in the Hume machine shop and equipped with automatic couplers and air brakes. The narrow-gauge log buggies were scrapped and 50 Russell skeleton-frame log cars were obtained second hand from the Itasco Lumber Company of Minneapolis, Minnesota, owners of the Minneapolis and Rainy River Railroad. The Humes first attempted to buy standard-gauge cars direct from the Russell Wheel and Foundry Company but the quoted price of $665 each seemed prohibitive. Thomas Hume, who had an interest in the Itasco Lumber Company, finally negotiated the sale of the second-hand cars at a price of $200 each. The cars, 23-feet, 10-inches overall length and in first-class condition, were shipped to Sanger on their own wheels and transported by freight wagon to Hume.

At the start of the 1915 season, use of the pine line of the railroad up Tenmile Creek was

Loading a redwood log at Camp Six about 1915. (Ira Stafford Collection)

discontinued completely as the two locomotives could not pull cars to and from the mill rapidly enough to permit logging in more than one area at a time. All efforts were henceforth concentrated on the switchback branch which had reached into the redwood groves by this time. Two new camps, Camps Four and Six, were established for redwood cutting, although a limited amount of fir and pine was also felled.

Rob Roy Hoist engine used at both Camp Four and Camp Six.

At Camp Four, an incline curved away from the railroad and dropped some 1500 feet into a big tree grove. Loaded cars were pulled up to the main line using the peripatetic Rob Roy hoist engine which had already seen similar service in Converse Basin.

Camp Six was near the foot of a long spur which branched away from the railroad main line and ran down a steep grade. Here, a small incline reached into an 80-acre timber tract called "Jones Park" after Bill Jones, the woods superintendent and later sheriff of Fresno County. Some "skyline" logging was also used to clear timber from the wooded slopes near Camp Six. In this unusual procedure, logs were swung on an overhead cable pulled by yarding engines, and carried high in the air across otherwise impassable terrain.

The total cut for 1915 improved slightly to 14,563,220 board feet. With the lumber market apparently on the rise, George Hume purchased a third Shay engine at a cost of $6435 plus freight. Additional log cars and several new yarding engines swelled the production potential. At this point, the company's logging equipment consisted of 70 railroad cars, three locomotives, 12 steam donkeys of various sizes, one McGiffert loader and one railroad crane. About 600 men were employed in the whole operation. In 1916, the Hume-Bennett seasonal output reached 19,184,380 board feet — the strongest showing in four

OPPOSITE: Building a railroad trestle on the line out toward Camp Seven. Note the interesting railroad crane.

Beautiful photograph by J. H. Traver shows the One Spot Shay on a long trestle above Camp Six. The portable camp buildings rest on skids below the railroad. (Harold G. Schutt Collection)

years but still far short of the break-even point.

A management change took place early in 1916 as a result of the death in February of George Hefferan, Hume-Bennett treasurer. His position was assumed by Thomas H. "Tim" Hume, the second of Thomas Hume's two sons. Young Tim was never very active in the California organization having other duties in Muskegon, but he delighted in kidding his older brother George about the costly enterprise whenever possible. He referred to the company on occasion by such terms as the "insatiable hunger," the "gigantic maw" and the "bank wrecker." George was called the "representative of the Golden West."

There had been discussion among the Humes for several years about the advisability of renaming the company. The "Bennett" portion of the title was not only an unbusinesslike misnomer but an unpleasant reminder of past association. On January 24, 1917, a corporate amendment was filed changing the name of the Hume-Bennett Lumber Company to "Sanger Lumber Company." (The new "Sanger Lumber Company" had no connection with the former and now defunct corporation known by the almost identical title of "*The* Sanger Lumber Company," the assets of which the Humes had purchased in 1905 as previously recounted.)

By 1917, the railroad had been extended farther into the redwood country and Camp Seven set up above Windy Gulch Grove near the end of the line. An incline was in use to bring timber out of a canyon in this area. The year 1917 more importantly marked the entry of the United

States of America as an active participant in World War One. The effect of this action was felt even in the far-away Sierra. The brawny and rugged Hume loggers made excellent prospective soldiers. Many quickly rushed to join the American Expeditionary Forces in a sudden burst of patriotism; others were very shortly drafted, and still others simply refused to work in the woods as the dearth of manpower opened up more lucrative jobs in less arduous fields. The company's ability to produce was seriously crippled by the eventual loss of about 40 per cent of its workers; shortages of supplies and equipment because of the wartime conditions increased the operating difficulties. To make matters worse, there was a flurry of local agitation at this very time by the Industrial Workers of the World, a socialistic labor organization. The I.W.W., as it was commonly called, was a militant protest group which opposed the policies of the craft unions of the American Federation of Labor. It was made up largely of immigrants, negroes and unskilled laborers of all types. The I.W.W. favored strikes and other direct actions by workers, and was not above bombings, arson or other terrorist acts to achieve its sometimes unreasonable demands. The I.W.W. agitators who invaded the Hume logging camps were very largely ignored by the average lumberjack, but George Hume became understandably concerned about the possibility of the company being the object of some violent act at the hands of the radical element. The situation became so worrisome at one point that special guards were hired to patrol the mill area.

Yarding donkey in action in the woods. The company had 12 steam donkeys of various sizes and styles.

Harassed by the multiplicity of troubles, the reduced Hume crew managed a cut of only 9,974,440 board feet of lumber in 1917. The war was uppermost in everyone's mind and the loggers followed the distant developments with patriotic enthusiasm.

The first motorized trucks used between Sanger and Hume Lake. It was a rough, bumpy trip at best.

ABOVE: Early day truck loaded with rails for the Hume railroad line prepares for the rugged mountain run. (Mac Grosse Collection). BELOW: One of the camp cook houses along the railroad east of Hume. (Vern Stafford Collection)

One wartime incident is worth recounting. A crane operator named Smokey Hughes managed to prevent several run-away railroad cars from crashing into the machine shop by derailing the wayward cars at the last moment. It was a noteworthy act but the crew very shortly tired of Smokey's apparently interminable bragging about the magnitude of his deed. It must have seemed to the men that Smokey thought his heroism had won the war singlehandedly. At any rate, a replica of a German Iron Cross, almost two feet long and very heavy, mysteriously appeared from the machine shop and was soon awarded to the surprised Hughes at an appropriate ceremony. The only trouble was that upon presentation, the burdensome cross was securely chained to Hughes' neck in such a manner that it could not be easily removed! It is reported that Hughes' wife, her husband's head resting on her lap, worked the entire night filing off the bulky encumbrance from her chagrined hero's person.

It is unlikely that Hughes ever again made mention of his feat — at least not in the presence of the Hume lumberjacks. When George Hume left California some years later, he remarked that the only item which he really cared to take with him was Smokey Hughes' Iron Cross!

The potential wreck averted by Hughes was one of few such incidents that took place on the Sanger Lumber Company railroad. Aside from occasional derailments caused by broken ties or spreading rails, there were no major difficulties on the line. The flume and mill operations were another story, however, and a substantial number of accidents and injuries occurred over the years. The flume itself, despite the most careful maintenance, was subject to frequent physical damage from fire, windstorm or lumber jam-ups.

On February 24, 1912, four miles of flume were lost as a result of a big windstorm. A fire later the same year burned several trestles along the route. On June 4, 1914, a severe storm blew down 96 boxes of flume causing a four-day delay in shipments. A lumber jam damaged 16 boxes of flume in August, 1914. In 1917, 26 boxes of flume were lost in a heavy rain, and a few months later, more than 60 boxes were knocked down by spilling lumber.

The flume workers had other worries besides that of keeping the big trough in continuous repair and operation. A man named Vandenbark missed his footing on the narrow flume walk on November 29, 1912, while working on the night shift. He fell 25 feet to the ground and was instantly killed although, strangely, no bones were broken. Giorgio Giraudo, a new employee, fell off the end of the tramway between the mill and the dry kilns on June 27, 1913, while pushing a loaded truck. He landed on the rocks beneath the platform and died soon after.

The log pond could also be dangerous. Bill Johnson, a loader, lost his ankle and foot in 1914 when he slipped and fell into the lake and became trapped. In 1912, a workman known only as "Ponzo" was raising sinkers from the lake by means of a hand-powered winch contrivance called a "mallard." He apparently fell into the water without anyone observing the occurrence, for at mealtime, when he failed to show up as expected, his absence went unexplained. It was not unusual for a disgruntled employee to take off for other territory at a moment's notice, so there was no great concern about Ponzo's whereabouts until it was noticed that he had failed to draw his pay

TOP: Redwood logs after bucking. CENTER: Loading a big log onto the railroad car at the bottom of one of the inclines east of Hume Lake. BOTTOM: This redwood log was so big, the front of the mill was cut away to permit passage. (Ross Phillips Collection)

ABOVE: A big steam yarder used to load railroad cars. (Mac Grosse Collection). BELOW: Interior of Hume cook house, Ira Stafford, long-time Sierra logger, in center.

and that his bedroll was intact. During the following winter, Jack Clark, Hume mill superintendent and winter caretaker at the time, asked permission to drain the lake and conduct a search of the bottom. The remains of the unfortunate Ponzo were soon found wedged between two sunken logs where he had obviously become trapped after falling into the water several months before. Some stories, perhaps apocryphal, say that his whiskers had grown considerably in the interim! Clark telephoned the Fresno County Coroner who made his way into the snow-covered mill area by horseback. After the death was officially certificated, Ponzo's corpse was lashed to a flume boat and sent on its lonely journey down the flume to the valley below where the gruesome load was picked up and properly buried.

Inside the big mill, workers sometimes carelessly allowed their clothing to be caught in the exposed machinery (safety regulations were not stringent in those days) and lost either an arm or leg, or had the appendages severely mangled. Many of these incidents resulted in drawn-out and expensive legal battles between the injured employee and the company or its insurance representatives. The gradual emergence of satisfactory compensation laws beginning about 1914, eventually improved conditions in this regard.

Excellent view of a flume trestle, 102 feet high, on the Tenmile Creek extension. (Harold G. Schutt Collection)

Loggers posed in their Sunday best in front of Anderson's saloon in 1912. (Jean Hume Browning Collection)

Deaths and injuries took place in the forest camps on occasion although conditions at Hume were not nearly so hazardous as they had been at the Converse Basin. The redwoods being felled at this time (seldom over 10 feet in diameter and usually much less) were mere toothpicks when compared to the monstrous, hard-to-handle Converse timber. Working in the woods was always dangerous, however, and a number of workers were maimed at various times by rolling logs or snapping donkey-engine cables.

But of all the trouble that beset the company during its stay in the Sierra, perhaps none proved more irritating than the relationship between the Hume management and a Swedish homesteader named Gustaf Anderson. In the early 1880's, Gust Anderson laid claim to 480 acres of Sierra forest land at the southwest corner of Long Meadow (later the site of Hume Lake as heretofore outlined). While Smith and Moore were operating their ill-fated Kings River Lumber Company at Millwood, Anderson sold them the timber rights to 160 acres of his land. This timber deed, although as yet unexercised, was subsequently absorbed by The Sanger Lumber Company when it succeeded Smith and Moore, and later, by Hume-Bennett as a part of its over-all purchase agreement. When he first came to California, Ira Bennett wanted to obtain all of Anderson's timber, and suggested to Thomas Hume that it might be advisable to buy up the two mortgages outstanding against the 480 acres and foreclose. Hume would not approve such a move as it was contrary to his principles to do business in such a fashion, and no further action was taken in this direction.

In 1906, Anderson obtained a liquor license and opened a log-cabin saloon on his property. The rustic tavern was patronized only by occasional hunters and trappers except for those Converse Basin lumberjacks who were willing to hike over a rudimentary trail to the Long Meadow oasis.

When the logging scene shifted to Hume in 1909, Anderson found himself with a ready made market for his wares and he proceeded to capitalize on the fortunate situation. Within a few years, he had built a summer resort of modest size which, in addition to the bar, included a store and several tent cabins for tourists. Much to the lumber company's dismay, the convenient saloon proved a great attraction to the thirsty Hume lumberjacks. On important holidays such as the Fourth of July, drunken celebrations sometimes extended over a period of several days or more, causing much valuable work time to be lost. Payday was another occasion for wild revelry. Many loggers cashed their checks at Anderson's bar, and usually spent a considerable sum for libation in the process. In an effort to forestall such occurrences, the lumber company itself began cashing employees' pay vouchers at its mountain office.

There was really very little that George Hume could do to alleviate the continuing Anderson problem. Hume Lake, at times of peak capacity, overflowed onto a portion of the saloon keeper's property, and Hume was fearful that legal action might result from this touchy situation if he antagonized Anderson unnecessarily. The affair soon resolved itself into a sort of armed truce, but Hume was never happy with the arrangement.

* * *

On October 30, 1917, the Hume mill closed down for the winter after suffering through a frustrating wartime season. Four days later, on November 3, just as the maintenance and repair crew of 60 men entered the cook house for lunch at 11:30 a.m., the pungent odor of smoke was noticed on the mountain air. Alarmed, the men rushed quickly back to the mill they had quit only moments before, and there found the entire front of the big wooden structure ablaze! Flames fanned by a strong up-canyon wind seemed to fill the sky and the frantic workers hurriedly set up fire lines in an attempt to extinguish the conflagration. George Hume, who was at the Sanger office when news of the fire reached him by telephone, jumped in his car and made it to the mill area in record time, arriving while the ashes were still smoldering.

Boiler house and engine house are all that remain of the Hume sawmill after the fire. Smoke is still evident.

Twisted wreckage of the dry kilns just after the big mill fire of 1917. (Jean Hume Browning Collection)

"The sprinklers were working," Hume later explained, "and we had steam on the fire pump, but water had no effect. The dry kilns and the unloader were doomed immediately and I cannot yet understand how the boys managed to save the yard and the machine shop as they were both afire in a great many places. We lost probably 50,000 feet of lumber piled on the tramway or near the mill."

When the holocaust had finally run its course, it left damages estimated at $500,000, only a part of which was covered by insurance. The big mill was destroyed, the dry kilns ruined and the upper portion of the flume badly wrecked. To prevent the flume from drying out, work was immediately begun to rebuild the burned-out section of the route between the dam and the end of the yard. This was completed in three days and shipments of lumber to Sanger were resumed at that time.

In view of the previous trouble with both the I.W.W. and Gust Anderson, many persons suspected arson as the source of the mill fire. Rumors that hinted darkly at sabotage made the rounds of the valley and there were even stories that the mill sprinklers were found to have been stuffed with rags. Both the local authorities and the insurance company conducted preliminary hearings on the matter but no incendiary cause for the blaze was ever established.

ABOVE: A big trestle on the Hume railroad heading out toward Camp Seven. (Ross Phillips Collection)

BELOW: Typical woods camp with portable houses where the loggers stayed while working in the forests.

While the fire was still a major topic of conversation, a new "cloak-and-dagger" event took place. Early in 1918, the Sanger Lumber Company powder house at Hume was broken into and a supply of dynamite stolen. In a short time, George Hume received a blackmail note which threatened dire happenings unless a certain sum of money were paid within three days. Hume, fearing the missing dynamite might be used to blow up the dam, and worried that harm might be done to his family, immediately hired two private detectives to investigate the source of the extortion attempt. He himself began carrying a revolver in the glove compartment of his car.

Suspicion for instigating the robbery and blackmail affair soon fell on one Yack Younger (or Yunger), an ex-lumber company employee who was then engaged as saloon manager at Anderson's resort. Younger made the applejack and whiskey sold at Anderson's (50 cents a bottle), tended bar and assisted with the "girls" who sometimes used the tavern as a location for hustling. According to the detectives' report, he appeared to be in a mentally disturbed state.

George Hume invited Younger to come to his house and discuss the evidence against him (Younger) in the hope of finding out the facts behind the blackmail attempt and possibly the fire as well. Hume, a small revolver hidden in his pocket, had the house empty of family and the duct to the furnace under the living room removed so that several witnesses below could record the conversation in legal fashion. The plot did not go according to plan. One of the listeners made too much noise and Younger became unduly alarmed. Moreover, the stenographer who was hired to transcribe the talk failed to work fast enough to get all that was said properly recorded. Younger, however, was arrested when he left the house by Sheriff Bill Jones and placed in jail. After a short trial, he was released on the grounds of insufficient evidence.

The strange case of extortion came to an end without any further developments and neither Younger nor the dynamite ever turned up at Hume again.

* * *

Limited repairs to the burned-out Hume Lake facilities were completed by the spring of 1918. The company moved in the remains of the machinery that had been used in the small Abbott Mill during Converse Basin days. By June, an open-air circular mill was in operation about 300 yards

TOP: Felling a big redwood near Camp Seven. BELOW: Two views of the circular mill built in 1918. (Both Jean Hume Browning Collection)

Stephen Mather-led party on a tour of the Hume facilities in July, 1919. At right is George Hume explaining layout to Mather and a ranger. (Bancroft Library)

up the lake from the former big mill. Because of wartime restrictions against new construction, there was very little market for lumber of any sort at this time, and the 1918 seasonal cut was only 6,518,705 board feet.

The Humes now faced a most perplexing dilemma. There was no point in continuing the lumber enterprise without constructing another big mill, and in view of the past financial history, such a costly outlay seemed foolhardy. Yet to allow the logging equipment and flume to stand idle was a wasteful expense, especially since the flume had to be kept in at least elementary repair if the company were ever to be sold as a going concern.

While the problem was still under advisement, a possible solution presented itself — the absorption of the Hume property into nearby Sequoia National Park, the enlargement of which was under consideration at this very time. Sequoia National Park and its companion, General Grant National Park, were created by an Act of Congress in 1890 as the result of efforts by early conservationists who wanted to preserve some of the outstanding groves of big trees for the public to enjoy. In 1915, Stephen T. Mather, first Director of the National Park Service, began a campaign to enlarge the park boundaries by the inclusion of the surrounding redwood country. During the next few years, Mather brought a series of influential groups to the Sierra in an attempt to obtain popular support for his proposal. In July, 1919, a Mather-led delegation took a railroad tour of some of the Hume holdings with George Hume himself acting as host and guide. A period of extensive correspondence about the matter followed the Mather visit and the proposition appeared promising. But like many political affairs, the issue dragged on inconclusively for several more years and finally died aborning. (When Sequoia National Park was ultimately enlarged in 1926, the newly acquired area did not include any Sanger Lumber Company land.)

* * *

On January 1, 1920, Thomas Hume died at the age of 71 after an illness of two years. As the eldest son, George Hume was suddenly burdened with much more extensive duties than that of manager of the California lumber enterprise. The elder Hume's holdings were large and varied, and required virtually full-time attention. Many companies, parts of corporations, and assorted land investments were included in the estate and it was some years before many of the various pieces could be legally apportioned.

While George was thus heavily occupied, the Sanger operation stood idle and no cutting was conducted for a three-year period. By the summer of 1922, Hume determined that it might be expedient at this time to log a stand of pine and fir which stood near the lake. The Pickering Lumber Company of Kansas City, Missouri, a nationally prominent concern, seemed seriously interested in the possible purchase of the Sanger Lumber Company and had, in fact, taken an option on the property and equipment. The flume was in need of repair after suffering fire damage earlier in the year, and Hume felt that the prospects for a sale would be advanced if the enterprise were in some sort of operation.

The small circular mill built in 1918 was refurbished. Crews were hired and the flume repaired at a cost of $100,000. In 1923, 10,152,288 board feet of lumber were shipped to Sanger by both flume and motor truck.

Part of the timber being cut came from Gust Anderson's 160-acre tract, the rights to which Smith and Moore had purchased many years before. A series of legal actions by Anderson held up logging of this timber until 1923, as Anderson claimed, without success, that the statute of limitations had expired on the original deal. Anderson later ineffectually sued the lumber company for $150,000, alleging that the beauty of his resort was marred by the 1923 logging.

Despite the renewed lumbering activity, the Pickering Company failed to exercise its option by the close of the 1923 season, and there seemed to be no point in continuing the hapless operation. There was brief hope that William Randolph Hearst, the famous publisher, might purchase the property after he inquired about "getting some redwoods for his grandchildren," but nothing concrete ever resulted from this overture.

In 1924, George Hume reluctantly took leave of his beloved California mountains and returned to Michigan. His family obligations as the titular head of the Hume fortune outweighing his personal desire for an outdoor life, Hume was virtually forced into the sedentary role of a respected business tycoon. Those who knew him in later life after ill health and financial responsibility had taken their toll, say he bore little resemblance to the gay and exciting young man who had so eagerly come west many years before.

The territory near Hume Lake is rugged ! This picture taken from above the railroad tracks looking north over the Kings River Canyon. (Jean Hume Browning Collection)

* * *

In July, 1926, a major forest fire destroyed more than 2000 boxes (16-foot sections) of flume in the rugged canyon along the Kings River. It was never rebuilt. From this time on, George

Hume and his fellow stockholders thought of only one thing in connection with the Sanger Lumber Company — the disposition of the land and equipment in the easiest and quickest way possible.

Buck McGee, a former company blacksmith, was appointed caretaker and watchman at Hume Lake, and Jack Brattin, a Sanger resident, assumed local responsibility for the business affairs of the company under direct orders from George Hume. McGee's principal task was to protect the mountain area from fires and vandalism, as the rugged country was an attraction for hunters, fishermen and hikers of varied skills and conscience.

On November 19, 1927, the physical equipment (not real property) of the company yards at Sanger, the flume from Hume to Sanger, and certain items at Hume itself, were sold for $20,000 to E. M. "Ned" Prescott, member of a prominent San Joaquin Valley family and a long-time Fresno area lumber dealer. Prescott also took a three-year option on the land at Sanger for an additional $10,000, and in the interim had free use of the property in question. Prescott was given extensive terms ($1000 down) on the acquisition and later, under a series of differing arrangements, purchased other portions of the equipment at Hume Lake.

Using the little circular mill, Prescott cut 649,393 board feet of lumber at Hume in 1929 (mostly sinkers and downed timber) and trucked it to Sanger for sale at his retail yard. At the same time, he marked off the flume into sections, each section priced according to how accessible it was, and sold the lumber to local farmers on an "as-is, where-is" basis. The flat-land sections of the flume from Sanger to Trimmer Springs were quickly disposed of, but the upper portion that wound through the mountains went begging. On July 3, 1931, a powder accident in connection with the building of Highway 180, set a fire in motion along Tenmile Creek which destroyed most of the remaining flume sections.

Prescott was a reputable operator but the 1929 depression nearly ruined him, as it did so many other businessmen. He was caught with an unsalable supply of lumber and equipment, and fell behind on his payments to Hume. Additionally, there was some dispute about the exact items to

ABOVE: Remarkable picture of the log-unloading trestle at Hume Lake complete with Shay's One and Two and a log train. (Annie Perry Collection). BELOW: Another fine view of the log-dumping area. The main line tracks in front were not curved as the camera distortion makes them appear. A shed covers the steam donkey in this picture. (Harold G. Schutt Collection)

Loggers pose beside a big undercut in this huge redwood tree. Although some large specimens were felled near Camps Six and Seven, none of these trees approached the Converse Basin giants in bulk. (Ross Phillips Collection)

which he had title, as some of the buildings and other exempted items were being disposed of concurrently by Jack Brattin, the company representative. A fire which destroyed the yards at Sanger further complicated the deal. It took more than 10 years for Prescott to finally clear up his obligations, but eventually nearly everything was sold, junked or otherwise removed at both Sanger and Hume Lake. Much of the scrap iron found its way to Japan and may well have been dropped in another form on Pearl Harbor in 1941.

While Prescott was dragging out his piecemeal physical cleanup, negotiations were taking place between the Hume interests and the United States Forest Service about the possible acquisition by the government of the mountain property owned by the Sanger Lumber Company. On April 8, 1935, after lengthy discussion, in excess of 20,000 acres of land including Hume Lake, Converse Basin and the surrounding territory, was sold to the Forest Service for a price of $319,276.75, less certain deductions, making an average price of $14.93 per acre over-all. The property ultimately became a part of Sequoia National Forest and belonged — after half a century — to all the people, just as it had in the very beginning.

The redwoods, or at least what was left of 11 redwood groves, had finally come full circle.

A load of railroad ties crosses a trestle near Tornado Creek. (Harold G. Schutt Collection)

IT WAS A LONG, LONG WAY...

FROM THE WOODS

TO THE RAILROAD LANDING

ALONG THE LINE TO THE LOG DUMP

INTO THE POND

UP THE RAMP INTO THE MILL

DOWN THE MIGHTY FLUME

TO THE SANGER YARDS

THE BIG GAME HUNTER

Joe Fourche, alias Joe Clark, was a French-Canadian who ran away from home at the age of nine. He later turned up as a lumberjack in the Hume area. An unusual character, he was rough, tough, big and burly — a true mountain man in every respect. He was an excellent marksman, an expert trapper and a clever hunter. His French name proved too hard for most people to pronounce so he took the name of Clark. He started a museum (below) at Hume Lake after attracting some attention with a pair of bear cubs which he captured and penned up for display at Converse Basin. In time he collected a large assortment of stuffed animals of all sorts and began charging admission to tour his museum. He won a prize at the San Francisco Worlds Fair for his excellent taxidermy. Next to his animals, he liked liquor best, and was constantly in trouble because of his escapades. Eventually, he was forced to move his museum to Pinehurst after some problems with the game warden and he died of pneumonia brought on by exposure to the elements following a drinking bout. (Both photos Jean Hume Browning Collection)

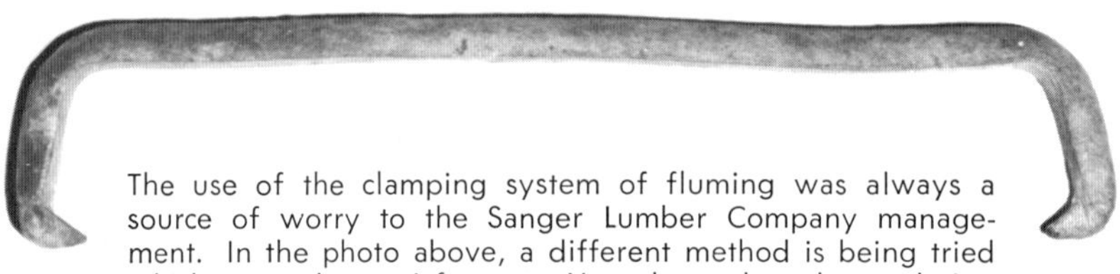

The use of the clamping system of fluming was always a source of worry to the Sanger Lumber Company management. In the photo above, a different method is being tried which proved unsatisfactory. Here loose boards are being received at Station Four, transferred to a clamping area (left), and then clamped in the usual fashion for the remainder of the journey to Sanger. At one time, "pegging" was considered. In this application, holes were bored in the lumber and a wooden peg then driven through the pile. Despite all the ideas and experiments, the original procedure was followed almost all of the time. This caption is framed by two pictures of flume clamps, each about 12 inches long in prototype size.

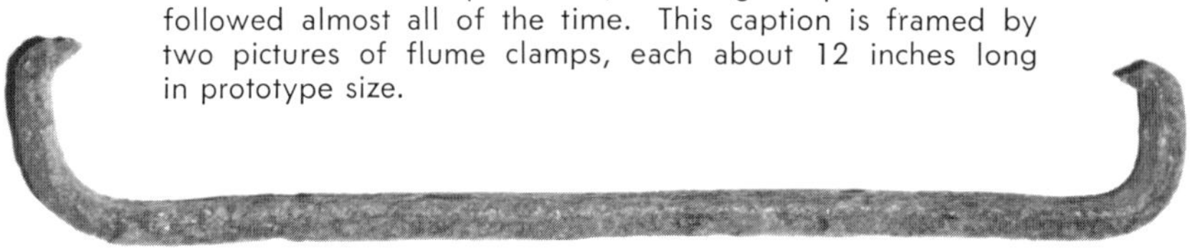

ABOVE: Here is photographic evidence of the reason no logging was conducted during the winter months !

BELOW: A musical flume herder relaxes on the porch of his flume house. (Walter Smith Collection)

Another trestle construction scene east of Hume Lake. (Ross Phillips Collection)

CHAPTER SIX

AND STILL THE BIG TREES GROW

DURING THE late 1930's, the Forest Service — working with Civilian Conservation Corps assistance — began a gradual and extensive cleanup of the obsolete lumber camps at Hume, Millwood and the Converse Basin. Work crews hauled tons of rusting scrap from the forests, dismantled and removed scores of decrepit buildings, and laboriously pulled hundreds of rotting stumps and sinkers from Hume Lake. In time, most of the logging residue had disappeared.

Today, unless one takes the difficult side trip into Converse Basin, it is not easy to tell that lumbering once threatened the very existence of the Sierra big trees. As the tourist drives easily over modern highways through Sequoia and Kings Canyon* National Parks, he sees a virtually unspoiled panorama unlike any other in all the world.

North of Highway 180, Sequoia Lake — the original Kings River Lumber Company reservoir — serves as a summer campground for the Young Men's Christian Association and the Boy Scouts of America. The YMCA, a San Joaquin Valley group operating under the name of the Sequoia Lake Conference, purchased the property in 1922 from George Hume, who personally owned it. The sale price was $30,000.

At Hume Lake, the mountain slopes are covered with excellent, second-growth forests. The lake is blue and beautiful. Visitors crowd the shores, and the public campground is often filled to capacity in season. The dam, although it has needed some repair, still stands rock-solid — a testimonial to John Eastwood's skill and originality. Aside from the concrete foundations of the big mill, there is little physical evidence of the old days.

Anderson's resort, operated in later years by Art Larsen, Anderson's nephew, was acquired in 1946 for $140,000 by the Hume Lake Conference, Inc., an interdenominational Christian association. A sizeable church camp has since been built on the grounds. It seems somehow incongruous that property with such an iniquitous early history is now the exact site of a place devoted primarily to the worship of God.

*Kings Canyon National Park was created by an Act of Congress in 1940. The former General Grant National Park is now a part of this park area.

Forest Service marker stands as mute evidence of the former lumbering operations in the basin.

East of Hume proper, a rough fire road follows the former railroad grade into the redwood country. Stumps are readily apparent in certain areas, but new growth is gradually filling in the cut-over lands. In 1950-51, the Forest Service split up the remains of nearly 1,000,000 board feet of redwood and sold it for fence posts. The timber had been left on the ground near Camp Seven following the mill fire of 1917.

* * *

George Hume, the last major figure in the Sanger Lumber Company story, died in 1944 at the age of 62. Before his death, he instituted negotiations for the sale of Mountain Home, the Sierra acreage acquired in 1907 by Thomas Hume, as previously recounted. It is to George Hume's eternal credit that no major logging was ever conducted in the Mountain Home Tract. For many years, he attempted without success to sell the land to various government agencies, and few could have blamed him for becoming disgusted with the continuing public apathy.

"We could get or work out more for the property than we are now asking," Hume wrote in a 1943 letter to his agent. "If the price we now suggest is considered too high, it is perfectly all right to turn down the property. My duty has been done, so far as attempting to preserve this timber is concerned; we have kept it for over 35 years without any cutting, but we cannot continue any longer. If it is not bought for preservation, the timber will have to be cut for lumber . . . I do not want to see this done.

"I am sorry now that we had to cut the sequoia from any of the Hume area, and I do not want to do it again in this Tulare district. Just as Smith and Moore are still censured for their old cuttings around Millwood and Converse Basin, I'll be cursed for a good many years for our cuttings east of Hume. I don't want such credit for still another district in California, when I agree myself that these magnificent trees should be preserved, to which I add, 'but not at private cost or expense.'

"At the price suggested," Hume concluded, "we will get back our costs and less than two per cent interest on the investment."

In 1946, two years after George Hume's death, the property was finally purchased for $550,000 by the California Division of Forestry. The Michigan Trust Company acted as agents in the transaction.

Mountain Home, containing four excellent big tree groves, is now a California State Forest.

* * *

The lonely whistle of the Shay locomotive no longer echoes across the Sierra meadows. The high-pitched whine of the big band saw has been forever stilled. The bucking steam donkeys, the rugged lumberjacks and even the mighty flume are all gone from view. The story of the lumbering of the Sierra big trees is essentially over. Except for isolated private holdings, nearly all of the giant redwoods are now safely in the hands of public ownership.

In looking back at our slipshod methods of awarding land grants, at the public apathy on the one hand and private greed on the other, it seems amazing, not that the redwoods *were* cut, but that the trees (and the American people) got off so lightly!

Today Hume Lake is a mountain playground.

ABOVE: Hank Johnston and Walter Smith standing on the last remaining railroad trestle of the Sanger Lumber Company in 1966. (Harold G. Schutt Photograph)

BELOW: At the left is an early picture of the Boole Tree in Converse Basin. At the right is Jean Hume Browning, a daughter of George Hume, pictured beside the old giant in 1966 — her first visit to the area since 1924 !

The ruins of Converse Basin are clearly evident in this photograph which shows only a very small part of the vast lumbering destruction. (Harold G. Schutt Collection)

The destruction of the Converse Basin was truly unforgivable. But had it not been for far-sighted conservationists — acting determinedly at a crucial period in our country's development — Giant Forest, Grant Grove and other irreplaceable areas might have also just as easily been lost.

Those who scoff at preserving our natural resources would do well to visit the Converse Basin. Two generations of shake makers have eliminated some of the original waste; a series of great fires down through the years has altered the appearance of the land, and incidentally, ruined much of the second growth, but the vast destruction is still clearly visible.

Walk through the meadows. Climb on the Brobdingnagian stumps — 20 to 25 feet across, 10 to 20 feet high in places. Their number is in the thousands, and they represent the relics of perhaps the mightiest forest that ever graced the earth.

It may well be that in some far-off millennium 1000 years hence, tall fluted columns of majestic redwoods will once again rise in profusion along the slopes of the Converse Basin. Nature has a way of healing even its severest scars. But until that time, and certainly within the life span of those who will read these words, the Converse Basin can serve only as a graphic reminder of the results of man's indifference and greed.

The very thought of it makes the heart ache.

Through all the wonderful eventful centuries since Christ's time — and long before that — God cared for these trees, saved them from drought, disease, avalanches, and a thousand straining leveling tempests and floods; but he cannot save them from fools—only Uncle Sam can do that.—John Muir.

BIBLIOGRAPHY

A LIST OF SOURCES

BOOKS

Adams, Kramer. *Logging Railroads of the West.* Seattle, 1961.

Andrews, Ralph. *Redwood Classic.* Seattle, 1958.

Bryant, Ralph. *Logging.* Boston, 1923.

Clark, Galen. *The Big Trees of California.* Yosemite, 1907.

Crump, Spencer. *Redwoods, Iron Horses and the Pacific.* Los Angeles, 1963.

Dalton, Emmett. *When the Daltons Rode.* Garden City, 1931.

Eaton, Edwin M. *Vintage Fresno.* Fresno, 1965.

Ellsworth, R. W. *The Giant Sequoia.* Oakland, 1924.

Evening News Association. *Men of Progress.* Detroit, 1900.

Farquhar, Francis. *History of the Sierra Nevada.* Berkeley, 1965.

Fry, Walter and White, J. R. *Big Trees.* Stanford, 1930.

Hanna-Kennedy. *Design of Dams.* New York, 1938.

Hutchings, J. M. *In the Heart of the Sierra.* Oakland, 1886.

Johnston, Hank. *The Railroad That Lighted Southern California.* Los Angeles, 1965.

Lewis Publishing Company. *History of Fresno County.* Chicago, 1891.

Manning, Rev. Samuel. *American Pictures.* London, 1886.

McKee, F. E. *Harnessing the Muskegon.* 1925.

Michigan Art Company. *Men of Michigan.* Detroit, 1904.

Moore, Charles. *History of Michigan.* Chicago, 1915.

Muir, John. *In American Fields and Forests.* Cambridge, 1909.

Muir, John. *Our National Parks.* New York, 1894.

Muir, John. *The Mountains of California.* New York, 1894.

Pillsbury and Ellsworth. *Business Directory of Tulare County.* Tulare, 1888.

Pinchot, Gifford. *A Primer of Forestry.* Washington, 1905.

Stewart, George W. *Big Trees of Giant Forest.* San Francisco, 1903.

Taylor, Norman. *The Ageless Relics.* New York, 1962.
Thayer, William. *Marvels of the New West.* Norwich, 1890.
Triplett, Col. Frank. *Conquering the Wilderness.* New York, 1883.
Vandor, Paul. *History of Fresno County.* 1919.
Western Publishing and Engraving Co. *Cyclopedia of Michigan.* Detroit, 1890.
White, John R. and Pusateri, Samuel J. *Sequoia and Kings Canyon National Parks.* Stanford, 1949.
Winchell, Lilbourne. *History of Fresno County.* Fresno, 1933.
World Book Encyclopedia. Chicago, 1957.

NEWSPAPERS AND PERIODICALS

American Lumberman (various issues)
Fresno *Bee* (various issues)
Fresno *Morning Republican* (various issues)
Fresno *Weekly Expositor* (various issues)
Holiday ("Redwood Epic." J. R. Challacombe. August, 1953)
Hutchings Illustrated California Magazine
Journal of Electricity, Power and Gas ("Hume Lake Dam." John Eastwood. October, 1909)
Los Tulares (Bulletins of the Tulare County Historical Society)
Muskegon *Chronicle* (various issues)
National Geographic (various issues)
Natural History ("Fire Ecology of the Giant Sequoias." Richard Hartesveldt. December, 1964)
Popular Mechanics ("When the Giants Fell." J. R. Challacome. June, 1954)
Porterville *Enterprise* (various issues)
San Francisco *Chronicle* (various issues)
Sanger *Herald* (various issues)
Scientific American ("Logging in the Sierra Nevada." December, 1896)
The Timberman (various issues)
Tulare County *News* (various issues)
Visalia *Times* (various issues)
Visalia *Times-Delta* (various issues)
Visalia *Weekly Delta* (various issues)
"Articles of Incorporation the Fresno Flume and Irrigation Company." Filed October 30, 1891, Fresno County Clerk.
"Articles of Incorporation the Fresno Flume and Lumber Company." Filed December 13, 1912, Fresno County Clerk.
"Articles of Incorporation the Hume-Bennett Lumber Company (A Michigan Association)." Filed December 14, 1905, Fresno County Clerk. (Amendment changing name to Sanger Lumber Company filed January 24, 1917.)
"Articles of Incorporation the Kings River Lumber Company." Filed April 18, 1888, Fresno County Clerk.
"Articles of Incorporation The Sanger Lumber Company." Filed October 15, 1894, Fresno County Clerk.
Corporate Records of the Secretary of State, State of California, Sacramento.
Legal Records of the Sanger Lumber Company.

DOCUMENTS, PAMPHLETS AND UNPUBLISHED MATERIAL

Anonymous. *The Redwoods of California.* State of California, Sacramento, 1964.
Anonymous. *The Status of Sequoia Gigantea in the Sierra Nevada.* Report to the California Legislature, Sacramento, 1952.
Davenport, Howard. *California Big Trees.* Calaveras Grove Association, Inc., Stockton, 1949.
Indenture Papers of Thomas Hume. County Tyrone, Ireland, 1862.
Letters From and To: Ira Bennett, George A. Hume, Thomas Hackley Hume, Thomas Hume.
McGee, Lizzie. *Mills of the Sequoias.* Unpublished Manuscript, 1952.
Sanger Lumber Company Business Papers: Journal, Ledger, Record Book, Stock Ledger, 18 letterbooks, 89 letterfiles, 37 Hackley and Hume Ltd. letterbooks, miscellaneous papers, blueprints, cutting records, abstracts, lawyers' opinions, inventories, articles of association, description of plants and yards, machinery orders, agreements, stock descriptions, shipping manifests, mortgages, bond issues and similar material.
Shirley, James C. *The Redwoods of Coast and Sierra.* Berkeley, 1936.

INDEX

Numbers in parentheses refer to illustrations.

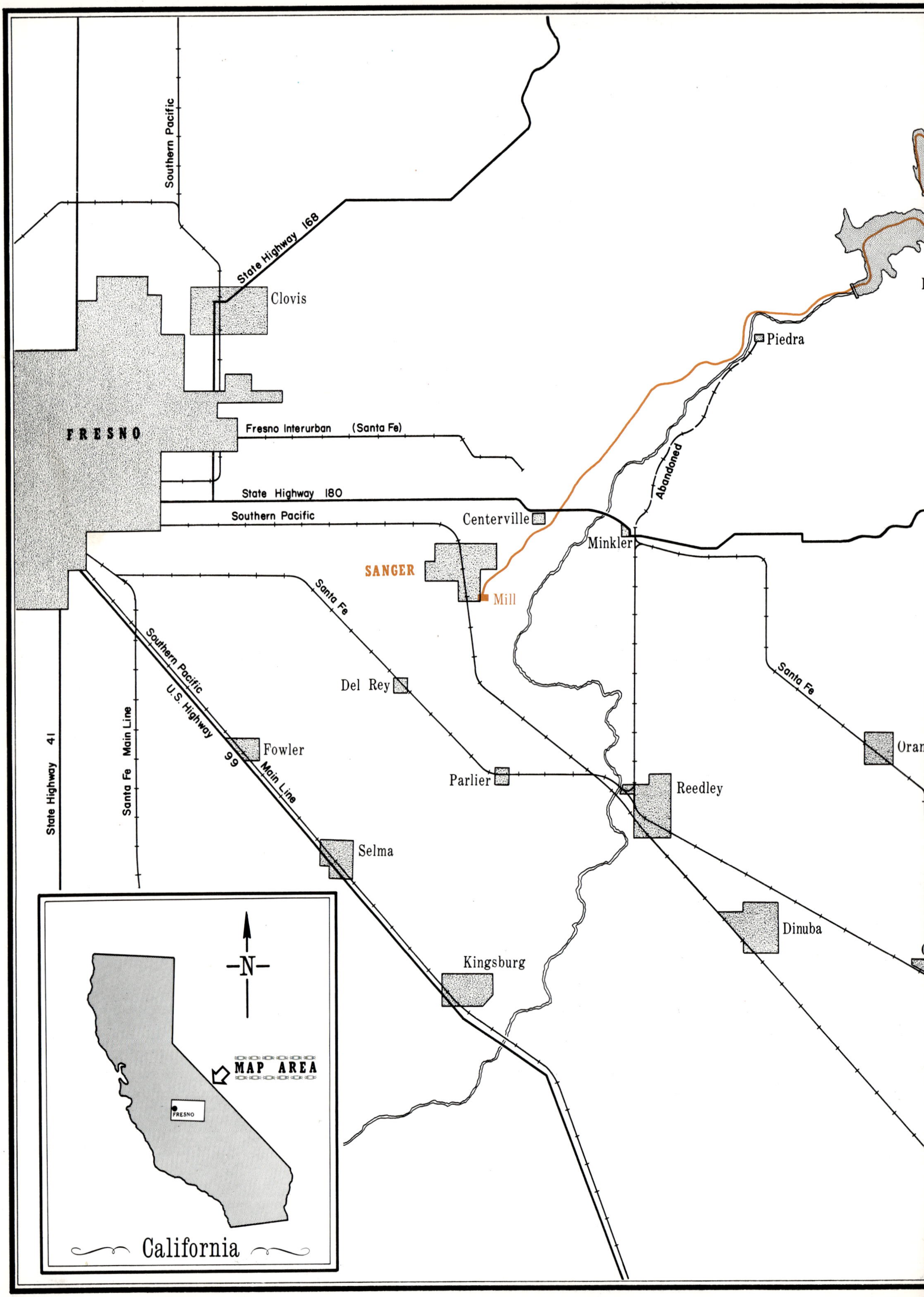

Southern Pacific
State Highway 168
Clovis
FRESNO
Fresno Interurban (Santa Fe)
State Highway 180
Southern Pacific
Centerville
Minkler
Abandoned
Piedra
SANGER
Mill
Santa Fe
Southern Pacific
U.S. Highway 99
Main Line
Del Rey
Fowler
Parlier
Reedley
Santa Fe
Selma
Dinuba
Kingsburg
State Highway 41
Santa Fe Main Line
N
MAP AREA
FRESNO
California